VAST EXPANSES

VAST
EXPANSES

A HISTORY OF
THE OCEANS

HELEN M. ROZWADOWSKI

REAKTION BOOKS

For Daniel

Published by
REAKTION BOOKS LTD
Unit 32, Waterside
44–48 Wharf Road
London N1 7UX, UK

www.reaktionbooks.co.uk

First published 2018
Copyright © Helen M. Rozwadowski 2018

Printed and bound in Malta
by Gutenberg Press Ltd

A catalogue record for this book is available from the British Library

ISBN 978 1 78023 997 2

Contents

The ocean surface: vast, trackless and opaque.

Introduction:
People and Oceans

Where are your monuments, your battles, martyrs?
Where is your tribal memory? Sirs,
in that grey vault. The sea. The sea
has locked them up. The sea is History.
– Derek Walcott, 'The Sea is History' (1979)

THE VAST EXPANSE of the world ocean, the dominant feature of planet Earth, has remained at the edges of our histories. Without conscious choice, writers have embedded a terrestrial bias in virtually all stories about the past. Dry land is the presumed norm. Even coasts and coastal dwellers have been viewed as marginal and exceptional, as have swamps, marshes, cays, reefs and other littoral areas that are neither entirely wet nor dry. Ocean basins appear on the fringes of land-based states and actions. Even events that took place at sea are often narrated as though the ocean is a flat, land-like plane without its underlying depths, having two dimensions instead of three. The time has come to put the ocean in the centre of some of our histories, not to replace terrestrial history but to add the history of the ocean itself to the other important histories we tell. Such a shift in perspective will yield rich dividends in our understanding of the past and equally enrich our present world in which ocean issues loom large.

This book attempts to tell the history of the ocean, extending its natural history to encompass its interrelationships with people, and including its depths as well as its surface. People have exploited the ocean for many reasons, starting with food and transportation, but also as the focus of myth and culture. New uses of the ocean emerged over time – science, communication, submarine warfare, mining and recreation. Alongside new uses, the old ones remain – piracy and naval warfare, shipping and smuggling, whaling and fishing. The ocean has enabled and constrained human activities, but people have also affected the ocean, in some cases dramatically.

Looking across its restless surface and smelling the salt spray, an observer today might imagine that past generations of seafarers or coastal dwellers witnessed the same ocean. Ships, breaching whales and even storms leave no tracks. Differences under the surface are not apparent. Yet the ocean is no less susceptible to natural and historical change than is the land. Just as the history of the land is inextricably intertwined with people, so too is the ocean's history, however hidden that history and however absent people have seemed from the sea itself. The perception of the ocean as timeless is as much a product of history as other cultural, political or economic changes resulting from the mutual relationship between people and ocean.

Far from providing a definitive telling of ocean history, this account offers a model, a starting point, in the hopes of inspiring others to embark on fuller, more inclusive histories. The goal here is to encompass all of the ocean, not just the slices along its coasts, surfaces or productive fishing grounds. The voluminous ocean provides over 99 per cent of the environmental space available for living things and, while people have not (at least so far) colonized the ocean, human activities make both tangible and imaginative use of all of the ocean, although clearly some parts more than others. This book charts a divergence of the history of a single global ocean appearing and changing over the lifetime of our planet into multiple histories of seas and coastal waters experienced by individual communities of people all over the globe. Since the fifteenth-century European discovery that all seas are connected, it is again possible to tell a story about the world ocean, as this book attempts to do. Hopefully, future histories will also investigate plural and individual oceans, defining and enumerating particular seas through physical geography, marine ecosystems, geopolitics, economics and also social and cultural conceptions.

The story related in this book weaves together three threads. First, the long story of the human relationship with the sea – all of it, including its third dimension – stretches to the scale of evolutionary time and extends over millennia to the present and future. Far from the ahistorical place it often seems to be, the ocean is profoundly a part of history. Second, the connections between people and oceans, though ancient, have tightened over time and multiplied with industrialization and globalization. Although we think of it as being starkly different, in this sense the ocean resembles the land. This trajectory runs counter to widespread cultural assumptions of the ocean as a place remote from and immune to human activity. Third, knowledge about the ocean – created through work and play, through scientific investigation and also through the ambitions people harboured for using the sea – has played a central role in mediating the human relationship with this vast, trackless and opaque place. Knowledge has helped people exploit marine resources, control ocean space, extend imperial or national power, and attempt to refashion the sea into a more tractable arena for human activity. Knowledge about the ocean, in short, has animated and strengthened connections between people and their oceans. Writing ocean history, I argue, must involve attention to questions of how, by whom and why knowledge about the ocean was created and used.

The first two chapters cover most of the time explored in this book, emphasizing the enormity of the ocean's past in the planet's history. 'A Long Sea Story', the first chapter, begins four billion years ago. It relates a deliberately ocean-centric story of Earth's development, in which dinosaurs appear in passing during the 'Age of Oysters', when molluscs dominated. Humans make their appearance as part of nature, within the natural history of the planet rather than separate from it, and ocean-oriented activities of early hominids and of *Homo sapiens* appear to have played an important role in the evolution of our species. 'Imagined Oceans', Chapter Two, continues the long story of people and oceans, finding that some cultures understood the sea as part of their worlds and territories, while others consciously turned away from the ocean. Before the fifteenth century, seas were known only locally or, at most, out to the boundaries of a usable basin. Specialist traders and navigators had experience of connections to an adjacent basin, but none knew the ocean as a global feature.

Although people across the globe lived by and with the sea for aeons, new knowledge of the ocean forged from the fifteenth through to the nineteenth centuries set new precedents for human use and perception of the sea, as chapters Three and Four explore. The era of geographic discovery by European powers, narrated in the third chapter, 'Seas Connect', etched water routes between all the Earth's known lands and laid the foundation for the doctrine of the freedom of the seas. Trade networks that criss-crossed the world provided the underpinning for an imperialism whose logic strongly promoted the exploitation of oceanic resources, especially the storied cod fisheries. Knowledge acquired through the work of navigation, warfare and fishing expanded with the Scientific Revolution to include discoveries made by practitioners of modern science that, in turn, enabled more intensive use of the ocean. Investigation of the vertical dimension of the seas, long a part of the work of some navigators and most fishers, began in earnest in the nineteenth century, as Chapter Four, 'Fathoming All the Ocean', recounts. New uses for the ocean, including the distant blue waters not previously exploited by people, expanded far beyond the traditional ones. The ocean transformed into a site for science, an industrial setting for transoceanic communications cables and a cultural reference that resonated with a generation fascinated by the sea.

Multiplication of new uses for the sea, alongside dramatic intensification of traditional maritime activities, has characterized the twentieth century, as chapters Five and Six show. Chapter Five, 'Industrial Ocean', examines the intensification of traditional maritime activities with industrialization. Expanding fisheries linked people with oceanic resources that were consumed far from where they were caught. Steam and iron quickened the tempo of development and also of everyday life, affecting the sea as much as the land. Beginning during the First World War, submarine warfare enmeshed the ocean's third dimension in global geopolitics. The Second World War involved unprecedented scientific investigation of the ocean to support undersea warfare, amphibious landings and sea-based aviation. In the wake of hostilities, the ocean emerged as a promising site for science- and technology-based economic development. As Chapter Six, 'Ocean Frontier', chronicles, inventors, entrepreneurs and officials transferred the metaphor of the American western frontier onto the sea to express their optimism for the growth potential of ocean-based industry. The scramble to claim oceanic resources led to the erosion of the centuries-long agreement regarding freedom of the seas. Extension of Exclusive Economic Zones may have ended the fiction of the ocean as a limitless frontier but did not appreciably curb intensive use of the sea.

As the final chapter and epilogue explain, a new posture towards the sea had its origins in post-war recreational access to the ocean. As Chapter Seven, 'Accessible Ocean', explains, the technology of scuba opened the undersea realm equally to frogmen, oil industry workers, scientists, casual divers, film-makers and others. The 1970s concern for the great whales and about the dangers posed by major oil spills drew attention seawards but did not translate into worry about the ocean itself, only its coasts and a handful of its more charismatic inhabitants. The accessible ocean, increasingly made visible through recreation as well as film, began a process of cultural transformation from robust frontier to fragile environment. Concern for the ocean as a whole gained traction only recently, however, as the Epilogue argues, with the belated awareness of overfishing and climate change and the extent to which these human interventions have remade the ocean.

The time to write ocean history is now. Recent scholarship from many fields has laid a promising foundation, revealing the underappreciated importance of the ocean and its depths, in both the past and the present. The fundamental quandary of the sea's apparent timelessness makes it difficult for us to accept the unfamiliar view of the ocean as a place of dynamic change. The humanities remind us that we know the ocean as much through imagination as through the knowledge systems of those who worked, or work, at sea. The opacity of the ocean guarantees that we see reflected back from its surface our fears and desires. Human motives, then, matter as much as biological interactions or chemical reactions. While present issues may seem to call for scientific and technological solutions, there remains a central and critical role for the humanities. Our understanding of the past will be revolutionized by an oceanic perspective that drives home the relevance of deep time and demonstrates the profound connectedness between people and the entire planet. The connection forged between people and oceans has changed both and tied their fates together. Our future may depend on acknowledging the ocean as part of – not outside of – history.

ONE

A Long Sea Story

All sea stories are true.

– Various mariner-storytellers[1]

POETS AND ORDINARY PEOPLE alike profess their love for the ocean, but the ocean does not love us back. It simply exists, although it does not exist simply, and it has done so since long before *Homo sapiens* evolved. Contrary to the tendency to think of it as a timeless, constant place, the ocean has changed dramatically over time. Throughout its mutable, four-billion-year lifespan thus far, it has played a leading role in nurturing life and fostering its diversity. As products of the profusion of life, humans were connected to the ocean first evolutionarily. Its natural history comprises, then, the earliest chapters of the story of the long human relationship with the ocean.

LIQUID WATER MAKES our blue planet distinct from all others in the solar system, making the formation of the ocean the first chapter, or at least the prologue, in our long story. Water became locked into Earth's earliest rocks as they took shape from dust particles in space to which water molecules adhered. On the early Earth, any water brought to the surface would have been released as steam, as rising temperatures melted rocks, and escaped from Earth because there was no atmosphere to trap it. Comets and asteroids carry water, and it appears that asteroids delivered the water that stayed on Earth as temperatures cooled and an atmosphere formed. Several cycles ensued of cooling, raining and continued asteroid bombardment that boiled off the water to produce dense steam, followed by rain once again as cooling continued.

From these processes, the ocean emerged about four billion years ago, only half a billion years after the planet itself took shape. At first, oceans covered most of the Earth's surface. Minerals that dissolved from submerged rocks and gases released by volcanoes and geysers entered the water, setting in motion the geochemical cycle that has kept the chemical composition of the ocean constant for a billion years.

In that primordial ocean, long before land emerged, rocks formed that provide evidence that life might have evolved and gained the capacity to photosynthesize by 3.8 billion years ago. Found now in southern Greenland and formed on an ancient sea floor, the Isua sediments are the oldest known rocks created at the planet's surface rather than deep in its interior. Actual microfossils of bacteria have been found in rocks dated to 3.5 billion years ago. The oldest ones were discovered in a Western Australian rock formation known as Apex Chert. This dark-grey, carbon-rich rock was laid down along the edges of a seaway near a volcano whose lava flowed over the seabed and sealed the fossils in place. Eleven kinds of thread-like microbes, some new to science and others indistinguishable from living cyanobacteria, reveal the ocean environment as host to morphologically diverse life extremely early in the Earth's history.

Until 3.9 billion years ago, our planet was bombarded with material from space, and a mere 65 million years ago a cosmic impact ended the age of dinosaurs. Any life that emerged early might easily have been destroyed, so that there may have been multiple life-starting events on Earth. Yet all life forms on Earth today are nearly chemically identical, and their roots trace to the same parental cell line. So it seems that one appearance of life took hold at an auspicious moment in the Earth's development. By contrast, prebiotic evolution of sorts happened on some asteroids but did not result in life, suggesting that water was critical.

While the evolution of life from non-life remains one of science's most enduring mysteries, one fact is known with confidence – that the ancient ocean played a major supporting role in this primordial drama. The most prominent spokesperson for evolution, Charles Darwin, recognized the centrality of a watery environment with characteristics different from the present Earth in his famous 'warm little pond' surmise, when he described to Joseph Hooker the conditions under which organic molecules might have given rise to a living organism.[2] His vision resembled that posited by scientists today: lagoons, lakes, puddles, groundwater and oceans enriched with organic compounds that, when exposed to atmospheric gases and stimulated by electricity, could produce molecules such as amino acids, sugars and other building blocks for life.

While we can be confident that the ocean served as life's cradle, there are several candidates for which part or parts of the ocean fostered this momentous innovation. Deep oceanic settings would have provided a refuge from cosmic bombardment. Areas at or near the sea floor would have available ferrous iron, dissolved from rocks, an essential catalyst for the synthesis of organic compounds. The discovery in 1977 of hydrothermal vents on the sea floor opened up a new possibility: that life evolved in proximity to deep-sea vents emitting hot water and gases. Vents likely served as a source of carbon for organic synthesis and, because they pump a volume equivalent to the world's oceans every ten million years or so, they regulated the chemical composition of the ocean.

For more than three billion years, life on Earth consisted of single cells or aggregations of cells that formed mats of microbes covering the sea floor. Not until some bacteria developed the ability to photosynthesize would the Earth's atmosphere gain oxygen. Accumulation of oxygen in the atmosphere, and eventually circulating throughout the seas, set the stage for multicellular organisms that could survive, and ultimately thrive, in the once-toxic soup of oxygen. Although earlier animal fossils have been found, those of the Cambrian period, starting about 540 million years ago, reveal a wild proliferation of life forms, all oceanic.

THE OCEAN'S ROLE in accommodating the stunning variety of life over the sequence of geological periods contributes a noteworthy series of chapters in our long story. The aptly termed Cambrian explosion generated the first representatives of many of today's taxonomic groups. The trilobites, early arthropods, dominated the period and spread over the globe. With their armoured external skeletons, these creatures swarmed throughout the warm, shallow seas for over 270 million years, filling a wide variety of ecological niches as predators, scavengers and plankton eaters. Other life forms included algae, invertebrates, echinoderms and molluscs, but not yet vertebrates nor terrestrial plants or animals. Adjacent to the diverse oceans, the land was relatively barren. Life did not yet exist in freshwater.

Much is known about Cambrian life thanks to the exceptional preservation of fossils, including their soft parts as well as their hard shells. In 1909, palaeontologist Charles Walcott discovered fossils in the Burgess Shale formation in the Canadian Rockies and dedicated his summers to collecting thousands of specimens. Decades later, scientists recognized the diversity and unfamiliarity of the fauna in his collection and made the Burgess Shale justifiably famous as a resource for studying evolution. Stephen Jay Gould's 1989 book, *Wonderful Life*, argued that Cambrian life displayed more diverse forms than exist on Earth today and posited that many of the unique lineages went extinct and, thus, represent evolutionary dead ends.

The spectacular fossil records of the Cambrian also record permanent changes to the sea floor. Competition for food in shallow seas promoted use of bottom sediments for avoiding predators and searching for food. Burrowing animals initially fed upon and were protected by the microbial mats covering the sea floor. Animals burrowing vertically began to break down the mats and make the upper layers of the sea floor softer and wetter. Their actions also allowed oxygen to penetrate below the sea floor's surface, irrevocably changing the environment at the bottom. In response, organisms dependent on microbial mats went extinct, to be replaced by new species adapted to the new conditions. Unfortunately for future palaeontologists, this dramatic environmental change also meant the end of conditions that permitted the exceptional preservation of fossils such as those in the Burgess Shale.

In the post-Cambrian era, life flourished in the stabilized, shallow marine environment. About 500 million years ago the first vertebrates appeared – eel-like creatures that lacked jaws and paired fins but sported a primitive backbone, head and tail. Unable to swim, they probably spent their lives wallowing on the muddy seabed, ingesting small food particles by filter feeding. At least two classes of jawless fish evolved, diversified and went extinct, or virtually so, during the Palaeozoic era (from 543 to 248 million years ago), but other fish classes that appeared during that time remain with us today. The lampreys and hagfishes in our seas have evolved a wide range of specialized lifestyles from parasitism to scavenging, to filter feeding. Their taxonomic status is disputed, but they might be descended from ancient classes of jawless fish.

Cartilaginous fish, including sharks, rays and skates, have skeletons made of cartilage, not bone, and further differ from bony fish in their lack of swim bladders and lungs. They pre-dated dinosaurs by about 200 million years and survived to fascinate us in the present. Sharks emerged 400 million years ago and proliferated in the Carboniferous period that followed. Some groups survived massive changes to the oceans, which caused several major extinction events and killed other marine life. Modern sharks first patrolled the seas as dinosaurs roamed but lived on after dinosaurs disappeared, numbering among the most ancient creatures on Earth today.

Bony fish, also still common in our seas, included a group that evolved into amphibians, which only late in the Palaeozoic era ventured onto land. Both ancient and modern amphibians remained closely associated with water. They had aquatic larval phases and required wet environments for laying eggs and for keeping adult skin moist. Reptiles evolved characteristics that finally broke life's ties to the sea. Hard-shelled eggs and scaly, dry skin retained moisture, freeing reptiles, and subsequently birds and mammals, to spread inland and across the globe to fill all types of environments.

Sea-floor spreading broke the supercontinent Pangaea into separate landmasses that began moving towards their current positions. Smaller divisions brought more land in contact with the ocean and created a sea between the African and Eurasian landmasses, which was christened the Tethys Ocean by the Austrian geologist Eduard Suess in 1893. Named after the sister and consort of Oceanus, the ancient Greek god of the ocean, and extant for 250 million years until destroyed by the movement of continents, the former Tethys exists today in the form of oil deposits in the Middle East and off West Africa and eastern South America that represent 60 to 70 per cent of the world's oil. Rocks in the Swiss Alps once lay in the western end of this lost sea, which extended in the late Cretaceous to cover what is now the Sahara Desert in North Africa, as well as vast parts of the North American Midwestern plains. Sea levels rose as the former Tethys Ocean floor bulged upwards, leaving only 18 per cent of the planet dry at the zenith of this episode of sea-level rise.

The Mesozoic era, or Age of Reptiles, was flanked by two extinction events. The largest mass extinction ever occurred 252 million years ago, eliminating 70 per cent of terrestrial species and over 90 per cent of marine species including the highly successful trilobites. As new life forms spread, dinosaurs dominated the planet for 135 million years. While schoolchildren everywhere know about the Age of Dinosaurs, few of us realize that this geological period could as accurately, though much less glamorously, be dubbed the 'Age of Oysters'.[3]

The oceanic food chain then, as now, depended on phytoplankton, primary producers that transform sunlight into food and provide sustenance for zooplankton. New microscopic plants and protozoans appeared, including coccoliths, diatoms, foraminifera and radiolarians, all of which produced shells, or tests, that contributed to different types of sea-floor sediments and chalks commonly found today. Molluscs proliferated, especially clams and snails that could burrow into the sea-floor sediment to escape the numerous predators in Mesozoic seas. The hard shells of oysters often proved insufficient protection from the powerful claws of crabs and lobsters, while starfish appeared whose suction feet were strong enough to pry shells apart. Nor were molluscs safe from predators who lived above the sea floor. Among the marine reptiles,

Henry de la Beche, *Duria Antiquior*, 1830: the first pictorial representation of deep time.

placodonts used broad teeth to crush shells of oysters and limpets. Some sharks and rays, even certain fish species, could defeat the armour of many molluscs, possibly to the point of causing the disappearance of entire species of bivalve brachiopods.

Open waters above the rough-and-tumble seabed were home to many species of cephalopods, a taxonomic group today represented only by squid, octopus and nautiloids. Coiled shelled ammonites evolved rapidly and spread widely throughout the seas. Many were likely good swimmers and formidable predators, with jaws capable of spearing or crushing prey, while others free-floated at various depths. Ammonites ranged in size from the diameter of a quarter-dollar coin to as much as 2 m (6½ ft), and filled many ecological niches. Their extreme abundance makes them excellent index fossils, helping geologists identify rock layers in which they are found, while their beauty attracts collectors.

Even the largest ammonites were dwarfed by the carnivorous marine reptiles that shared their seas. Discovery of large fossils of plesiosaurs or ichthyosaurs in the nineteenth century fascinated the public and encouraged reputable scientists to consider seriously that sea monsters might still exist. Icthyosaurs, striking examples of

Popular literature often featured illustrations such as this one from 1874, of a battling ichthyosaur and a plesiosaurus, by French illustrator Édouard Riou.

convergent evolution for their similarity in form to dolphins, were admirably adapted to Mesozoic seas, with well-developed paddles for locomotion and long bills with teeth for catching fish. As today's marine mammals did, they evolved from land-based species. While most ichthyosaurs were only 3 to 5 m (approx. 10 to 15 ft) long, some were 15 m (49 ft). Plesiosaurs had whale-like bodies, short tails and paddle-shaped fins. Long-necked forms that resemble the cartoon image of Nessie, the Loch Ness monster, are most familiar in popular culture. The smallest species measured around 2 m (6½ ft), while the largest could be 20 m (65½ ft) long, as large as today's sperm whales. These big marine predators pursued fish, sharks, ichthyosaurs, dinosaurs and other plesiosaurs.

The Mesozoic chapter closed dramatically with a mass extinction best known for killing off the dinosaurs, sparing only species that would evolve into modern birds. Possibly set in motion by an asteroid impact about 66 million years ago that left the Chicxulub Crater on the Yucatán Peninsula, 50 per cent of all genera disappeared globally in a short

Fossil skeleton of an early form of whale in Wadi Al-Hitan, the Valley of the Whales, a palaeontological site southwest of Cairo, Egypt.

geological period, including the ichthyosaurs and plesiosaurs that had ruled the seas, and all the ammonites, but also many species of micro-plankton, brachiopods, fish and land plants. In place of the large reptiles, birds and mammals diversified and remarkably rapidly filled every available ecological niche, including in the seas. Our own geological era, the Cenozoic, witnessed the evolution of land-based species of mammals into marine creatures.

Why did mammals return to the sea? Seafood might be the answer. Marine mammals first appeared as global temperatures and primary productivity in the oceans increased during the Eocene, and all groups evolved adaptations for feeding in water. As the Eocene ended and the Tethys Sea closed, part of the Southern Ocean opened when Australia moved northwards away from Antarctica. The drift of South America away from Antarctica opened the rest of the Southern Ocean, allowing the circumpolar current to flow. Connections between oceans resulted in mixing of ocean water, which increased productivity.

Fifty million years ago, ancestors of manatees, or sea cows, and cetaceans appeared in riverine and nearshore environments. By 40 million years ago, fully oceanic forms inhabited seas in southern Eurasia and the western tropical Caribbean. A UNESCO World Natural Heritage site in an Egyptian desert presents an unexpected glimpse of the ocean's past embedded in windswept sand and rock. Wadi Al-Hitan, or Valley of the Whales, holds hundreds of skeletons of marine mammals in the final stage of losing hind limbs. These reside alongside skeletons of sharks, bony fish, crocodiles and sea turtles deposited between 40 and 39 million years ago and preserve evidence of a shallow marine habitat which appears, from the large number of young marine mammal specimens, to have been a calving ground.

THE SPECTACULAR DIVERSITY of life fostered by the ocean responded in some instances to its extremes and in others to its more ordinary settings. Life evolved, and endures, unequally throughout the sea and uniquely in response to the extraordinarily varied conditions that form the subject of another chapter of the long story.

Fossils of marine mammals can be found on every continent, from the poles to the tropics, although they are most common in northern temperate regions. Today's marine mammals have a global but patchy distribution. Nutrients reach the surface through upwelling, a process that brings cold, dense, nutrient-laden water upwards when wind-driven circulation patterns push warm surface water away from coasts. Upwelling zones, usually on the western edges of continents or along seamounts or coral reefs, attract prolific marine life including fish and mammals. The daily vertical migration of zooplankton such as krill, to feed on the surface at night but avoid sunlight during the day, attracts other marine life to such zones.

Patchiness is a general characteristic of the ocean, as of land. That the sea covers 71 per cent, or almost three-quarters, of the surface of the planet, is a familiar fact. Less familiar is the statistic that the ocean comprises 99 per cent of the habitable space on Earth. It's a staggering thought. It means that the varied environments contained within the sea offer vastly more potential habitat than the immense mountain ranges, the sprawling rainforests and the wide plains and deserts of the Earth put together. But life is far from equally distributed throughout the ocean. Whale falls, cetacean carcasses that drop to areas about 2,000 m (6,560 ft) deep, provide resources for opportunistic deep-sea organisms in a localized patch of ocean floor for decades, and collectively play a crucial role in supporting the deep ocean floor ecosystem. Many ocean-ic life forms cluster in highly productive places such as richly prolific offshore shallow banks; cold, fertile upwelling zones; plankton-laden circumpolar open seas around Antarctica; waters close to features such as reefs; shelf breaks or seamounts; and the mysterious hydrothermal vents in the dark depths.

The discovery of hydrothermal vents in 1977 in a zone where the sea floor is actively spreading paved the way for the more startling encounter with strange, highly specialized communities of creatures living near active vents. Water percolates through cracks in sea-bottom spreading zones, gets heated and chemically modified by encountering hot rocks, and shoots out of available channels as hot springs, mixing with dissolved minerals in seawater. The first vent field found with life, poetically christened Rose Garden, was covered with giant tubeworms over 2 m (6 ft) tall. Persistent springs initially attract crabs, which subsist on bacterial mats. Then colonies develop of tubeworms, mussels, clams, crustaceans, specialized worms, or other invertebrates. These animals live within millimetres of superheated water, as hot as 350°C, because the pressure at great depths prevents hot water from turning to steam. The microbes at the base of these food chains do not, as elsewhere on our planet, rely on sunlight; they instead use hydrogen sulphide from vent fluid to create simple sugars. Some larger animals, such as tubeworms and mussels, depend on symbiosis with such microbes for their food. The productivity of vent communities rivals that of shallow-water coral reefs or salt marshes, but communities appear and disappear as the vents turn on and off.

Hydrothermal vents, with their incredibly steep temperature gradients, number among the many examples of oceanic extremes. Utterly unbeknownst to the Romantic thinkers of the nineteenth century who embraced the sublimity of extremes in terrestrial nature, the ocean hides the globe's largest mountains, the most gigantic volcanoes and the tallest waterfalls. The deepest part of the ocean, the Mariana Trench (which reaches 10,994 m, or 36,070 ft), would drown Mount Everest (which is 8,848 m, or 29,000 ft, tall) and extends more than five times the length of the 446-km (227-mi.) Grand Canyon, stretching to 2,550 km (1,580 mi.). The mid-ocean ridge, whose 72,000 km (44,739 mi.) scar the sea floor all around the Earth, covering 28 per cent of it, is the single largest geological feature on the planet's surface.

Oceanic extremes are not limited to geology. Fossils reveal former giants that patrolled ancient oceans, such as the Megalodon shark, at 14 to 18 m (46 to 59 ft) one of the largest and most powerful vertebrate predators ever. A relative of today's sperm whales and similar in size to

Megalodon, the giant carnivorous whale *Livyatan melvillei* had massive teeth in both the upper and lower jaws that it used to tear off flesh from its prey, probably baleen whales smaller than today's humpback whales. Cretaceous seas were also home to bivalves with shells over 2 m (6½ ft) in diameter and marine turtles almost twice that size. Today's blue whale remains the largest creature ever to live on Earth, growing up to 30 m (just under 100 ft) long.

The ocean may also hold the planet's oldest creatures. Deep-sea corals, which prosper without sunlight in deep, cold water, are the oldest marine organisms on Earth. One individual colony found off the Hawaiian coast was determined to be about 2,700 years old, while another individual colony of a different species had lived over 4,000 years. The Greenland shark is the longest-lived vertebrate known, growing less than 1 cm (½ in.) per year, reaching sexual maturity around 150 years of age and able to live perhaps up to four centuries. Arctic bowhead whales have been caught in the twenty-first century with harpoon points dated to the late nineteenth century, bolstering traditional indigenous knowledge asserting that bowheads could live two human lifetimes. Modern analysis of amino acids in whale eyeballs estimates lifespans of up to two hundred years. By contrast, elephants can live seventy years and occasional humans to about 100.

Oceanic life not only evolved to inhabit all possible niches including the extremes, but it developed behaviours to take advantage of regular changes in the ocean environments. The vertical migration of zooplankton, for example, allows them to follow darkness to escape predators by day and feed at night. Seasonal shifts in water circulation and temperature affect distribution of food resources, and marine populations have learned to follow. North Pacific humpback whales, for instance, migrate from cold, fertile northern feeding grounds off Alaska to breeding grounds in the warmer waters around Hawaii, while separate populations of southern humpbacks travel between Antarctic feeding grounds and tropical seas. They do not mix with their northern counterparts because seasons are reversed in the two hemispheres. Humpbacks

Left Gold coral (*Gerardia*).
Below Black coral (Leiopathes); one specimen of *Leiopathes* was found to be over 4,000 years old.

probably learned to migrate in response to an evolutionary carrot-and-stick combination of seeking sufficient food and avoiding predation on calves. Another very long whale migration, possibly the longest, is the 16,000- to 23,500-km (10,000- to 14,000-mi.) trip of the eastern Pacific grey whale to and from the Bering, Chukchi and Okhotsk seas from and to the Baja lagoons and the coast of California.

Other lengthy oceanic migrations are similarly tied to reproduction, including those of salmon, sea turtles and eels. Sea turtles, which must find beaches to nest and lay eggs, can swim 3,200 km (2,000 mi.) to do so, as green turtles do between the coast of Brazil and Ascension Island. Leatherback turtles have been found 4,000 km (2,485 mi.) from nesting beaches, presumably in pursuit of the jellyfish they eat. European eels presented a mystery that remained unsolved until the early twentieth century: young eels entered rivers and adults left, but no one knew where they went to spawn. In a series of expeditions between 1904 and 1922, the Danish marine biologist Johannes Schmidt located smaller, younger eels the farther he travelled out to sea, piecing together the life history of the eel. The larvae float along the Gulf Stream to European coasts, growing over one to three years into transparent eels that move into inland waters where they feed and grow for a decade or more before maturation propels them on the 6,000-km (3,700-mi.) open ocean journey back to the Sargasso Sea of their birth to spawn.

The life-cycle journey of wild salmon is the mirror image of the eels' travels. Hatchlings grow up for one to three years in rivers surrounding the Atlantic and Pacific oceans and then migrate to the ocean to feed and grow to maturity before returning to their natal river to spawn. Knowledge of their oceanic wanderings was virtually non-existent until fisheries developed in open sea areas in the post-Second World War period, touching off competition for the resource that motivated tagging studies which proved that the targets of these fisheries did, indeed, return to rivers where existing fisheries fully utilized the stocks.

OCEANIC ANIMALS ARE, of course, not the only ones that undertake long migrations. Terrestrial creatures such as Monarch butterflies, Canada geese and caribou do as well, and the longest migration known is the 35,000-km (21,750-mi.) journey of the Arctic tern to Antarctica. Hominids numbered among the species which migrated seasonally to take advantage of food and other resources and also dispersed to colonize varied new environments throughout the world. When human ancestors evolved, they did so in a world that included ocean as well as land. Movement along, around and over the ocean begins the earliest chapter of the long story of hominid and human relationships with the sea, which features use of the ocean for food and other resources as well as for transportation.

Archaeologists have long understood *Homo sapiens* and its ancestors as fundamentally terrestrial for most of evolutionary history. Until relatively recently, most scholars believed that intensive seafaring and fishing societies only appeared about 10,000 years ago, or less than 1 per cent of the time our genus has been on Earth. New scholarship in archaeology and historical ecology has dramatically extended awareness of the degree to which people have relied on the ocean. Archaeologists and other scholars are discovering earlier evidence of humans voyaging across deep water and along coasts and relying on marine resources to survive and thrive.

Global human migration depended on the ocean longer ago than scholars have previously believed. Compelling evidence comes from the tiny island of Flores, which lies east of the Indonesian island of Java. A hominid species arrived there roughly one million years ago, in the era known as the Early Pleistocene. To offer some perspective, *Homo sapiens* did not colonize this region until between 100,000 and 50,000 years ago. The fossils discovered at Flores so far reveal hominids similar in body and cranial size to australopithecine (an earlier hominid thought to have been confined to Africa). Archaeologists have concluded that their fossil discoveries represent a new species, *Homo florisiensis.*

Experts are not sure whether the immigrants to Flores were *Homo erectus*, which subsequently evolved under selection pressure in a calorie-poor environment for small body size, or whether they might have been an unknown, small-bodied, small-brained hominid. Either way, the implications for our understanding of prehistory are staggering. Archaeologists wonder at the thought that we shared our planet with other hominids much more recently than we have believed, including at about the time that agriculture was first invented, and long after Neanderthals became extinct. This discovery also opens up new chapters of ocean history.

Even at times of low sea levels, travel to Flores would have required the crossing of a deep-water sea barrier of 20 to 30 km (12 to 18 mi.), a crossing beyond the dispersal abilities of most land animals. Indeed, the only ones to colonize Flores without human assistance were rodents, which presumably arrived on natural rafts or other flotsam, and Stegodon, ancestors of elephants, which had the capability to swim there. The presence of these hominids on this otherwise inaccessible island provides indirect evidence of the oldest human maritime voyaging anywhere in the world. Previously, scholars thought that the organizational and linguistic ability to undertake sea voyaging appeared first in modern humans and only late in the Pleistocene. Some archaeologists insist that earlier crossings must have been accidental, but evidence is mounting in favour of intentional Pleistocene seafaring. The Flores findings also suggest that more groups dispersed out of Africa earlier than experts previously believed.

The span to be crossed to reach Flores was, in fact, part of the most pronounced biogeographic division on our planet, known as the Wallace Line in recognition of Alfred Russel Wallace, the nineteenth-century British naturalist and explorer better known as the co-discoverer of the theory of evolution by natural selection. In his travels throughout Indonesia he noticed stark differences between the northwestern and southeastern parts of the archipelago. On the Eurasian side, in places like Sumatra and Java, the mammals disappeared, seemingly replaced on the Australian side by such zoological oddities as marsupials and giant lizards such as the Komodo dragon. Such stark differences across similar climate and terrain presented a puzzle until Wallace recognized that the barrier was oceanic, not terrestrial – a swathe of ocean had kept Eurasia and Australia separated since the Last Glacial Maximum about 21,000 years ago. Scientists studying coral reefs in Indonesia today find evidence that the distribution of marine species is similarly influenced by a marine version of the Wallace Line.

Between the islands of Lombok and Bali, today 193 km (120 mi.) apart, there exists a strip of extremely deep ocean that remained a seaway throughout the movements of landmasses and the cycles of sea-level drops that allowed animals to migrate between many places that are now islands or separated landmasses. Few terrestrial mammals managed to cross the Wallace Line – just rats and people – until hominids deliberately transported dogs, pigs, macaques and other animals. Even terrestrial animals that swim well, including pigs, hippos and deer, did not cross unaided.

Evidence from Flores and along the Wallace Line suggesting that Pleistocene seafaring began at least one million years ago gained support from a surprising source. A finger bone from a young girl found in a cave north of Mongolia in Siberia set in motion a series of genetic studies that may connect her people to the sea. Based on the DNA in the bone, scientists have identified a previously unknown group of hominids named the Denisovians – the first example of such a discovery resting on genetics instead of anatomical description. More closely related to Neanderthals than modern humans, the Denisovians became a separate group sometime between a million years ago and perhaps 400,000 years ago. Evidence that they interbred with modern humans has been found through genetic comparisons with indigenous populations in Australia, New Guinea and surrounding areas. Surprisingly, the mysterious girl appears unrelated to people around the Siberian cave harbouring the only existing fossil specimens of her people. This distribution prompts some scholars to ponder whether the Denisovians themselves might have crossed the Wallace Line, or indeed whether *Homo florisiensis* was a Denisovian.

Hominids before *Homo sapiens* probably used the sea for travel, and may have achieved significant sea crossings. Members of our own species certainly engaged in ocean travel, initially, so anthropologists now think, moving along the coasts of the Indian Ocean. Genetic evidence suggests that people reached places in Melanesia and Australia that required sea crossings earlier than central Europe or inland Asia. Yet evidence is sparse, so mainstream archaeology has been slow to consider seriously early human migration by water. One problem is posed by today's sea level, which is more than 90 m (300 ft) higher than it was 18,000 years ago. Past ice ages, including the most recent one starting 50,000 years ago, locked up seawater and created habitable land, now inundated, that might have provided clues. Most knowledge of prehistory comes from inland sites. Underwater archaeology now possesses the tools and knowledge to extract information about human habitation from formerly dry land that is presently under the sea, although the expense of ship time compounded by continued terrestrial bias has kept this emerging field small.

Finds illuminating ancient human settlements come from fortuitous discoveries in rare places that were long ago close to the sea and still remain so, such as caves high up on steep cliffs. One such place is a high-elevation cave near Mossel Bay, South Africa. Evidence from the Pinnacle Point cave revealed that a group of humans there used shellfish intensively at a critical time, 164,000 years ago, when harsh environmental conditions caused by glaciation may have forced them to survive on food resources from the sea.

Marine food sources, which archaeologists had long believed early humans only recently exploited in evolutionary terms, are now recognized as having been important much earlier, and during at least one critical juncture for humanity's survival. It is well known that *Homo sapiens* likely emerged about 200,000 years ago in Africa. A glacial period stretching from about that time until 125,000 years ago created cold, dry conditions, making terrestrial food sources less productive and perhaps driving early humans to find or depend more heavily on new food sources. The Pinnacle Point cave findings suggest that a small population of modern humans may have hunkered down by the coast and expanded their diet to include shellfish. In South Africa, archaeological evidence indicates that this dietary dependency on shellfish coincided with symbolic behaviour and technological developments that together suggest these early humans displayed key elements of modern behaviour. The time and place of this find corresponds to a bottleneck in human evolution, making it possible that the inhabitants of Pinnacle Point cave might just possibly have been ancestors to all modern humans.

There is no need to rely entirely on the drama of an evolutionary bottleneck, however. Recent scholarship concludes that wherever aquatic resources were abundant and relatively accessible, our ancestors have likely always used them, probably longer ago than at Pinnacle Point. Neanderthals ate molluscs and also exploited seals, dolphins and fish during the Middle Palaeolithic period (200,000 to 40,000 years ago). Evidence from multiple sites suggests that such resource use was regular and sustained, a part of purposeful visits to coastal and estuarine environments. Agriculture emerged only very recently by comparison, about 10,000 years ago. For most of human history, foraging and hunting provided sustenance – and food came from both land and sea.

Reaching back further in time, hominids pre-dating *Homo erectus* also used aquatic food as far back as two million years ago. Archaeological investigation in northern Kenya found evidence of the butchering of turtles, crocodiles and fish from wetlands and the coasts of rivers and lakes of the East African Great Rift Valley. This area, unlike prehistoric savannahs or forests, offered a year-round food supply that was easy to access. Because these aquatic foods are rich sources of nutrients, particularly fatty acids (needed in human brain growth), it is possible that human evolution may have been spurred when hominids began eating fish and other such foods.

A theory about an aquatic phase during human evolution which faced intense scepticism for decades has garnered more serious, if still cautious, consideration in recent years. The so-called 'aquatic ape' hypothesis is credited to the highly regarded marine scientist Alister C. Hardy who revealed his thoughts, apparently kept hidden for decades, in 1960, three years after he was knighted. His theory rested on the observation that apes can walk upright but don't unless wading through water (or carrying fruits or sticks) and posited adaptations such as human hairlessness and the relatively large amount of subcutaneous fat as vestiges of a semiaquatic lifestyle, in which early humans inhabited swampy or marshy areas to escape predators and dove to shallow depths to get food. In the 1970s, the feminist author Elaine Morgan questioned the scientific basis for theories asserting that hunting by men drove human evolution. She promoted the aquatic ape hypothesis as an alternative that would account for contributions by women to subsistence gathering as a force for evolutionary change.

Regardless of a possible aquatic phase in hominid development, the bare facts of human use of the ocean for migration and food suggest that use of the ocean has not been well integrated into our understanding of prehistory. Coastal living allowed for the exploitation of inland as well as coastal resources, which together could sustain a community year-round. The ease of shellfish gathering meant that women, children and the elderly could contribute regularly and significantly to subsistence. Plentiful food may have provided the time and opportunity for humans to develop tools and crafts, to build communal structures or possibly to experiment with cultivating plants. Access to water likely promoted communication and trade with other groups, perhaps ultimately stimulating dispersal along coasts and around the globe. Much of the story of ancient human relationships with the ocean remains unwritten.

THE CHAPTER CHRONICLING human migration to North and South America exists as an old familiar tale undergoing active revision as archaeologists learn more. Did the first *Homo sapiens* to set foot in the Americas arrive after an arduous trek across the Bering land bridge and through a long 'ice-free corridor' soon after glacial ice lost its grip, devoting generations to reach North America and more to expand southward? Or did they put their possessions in boats and skip along the coast, pulling up on the shores of islands or favourable beaches for long or short pauses? At glacial periods with lower sea levels than today, very broad coastal zones provided natural highways for human exploration and dispersal, entirely at sea level and with very few geographic barriers. Shore environments up to 200 km (124 mi.) wide offered varied terrestrial and marine food sources, relatively flat terrain and possibly convenient beach access for watercraft that may have speeded travel.

The old theory is fraying that the Bering land bridge and ice-free corridor were the keys to the peopling of North America. Evidence is mounting of older settlements, whose inhabitants could only have arrived by sea. The people who first settled the Americas were long believed to be big-game hunters from Siberia and Beringia who traversed the ice-free corridor as the Laurentide Ice Sheet retreated, about 13,000 years ago. Archaeologists have found numerous sites characterized by tools of the Clovis culture, named after the Clovis, New Mexico, location where they were first discovered, and evidence that the arrival of these tools' users coincided with the disappearance of big game. Even as the Clovis first theory gained acceptance, though, anomalous finds of likely older evidence of human habitation raised questions.

Monte Verde, Chile, lies thousands of kilometres south of Beringia. There archaeologists discovered artefacts dated to between 15,000 and 14,000 years ago. This corresponds to the time that the ice-free corridor first opened, but it may not have been biologically viable until a couple of thousand years later (13,000 to 12,000 years ago). The artefacts do not resemble Clovis-type tools, and there was not sufficient time for people to move 16,000 km (10,000 mi.) to create the settlement found there. Other sites pre-date Clovis technology and the opening of land access to the Americas, including one at Paisley Five Mile Point in Oregon, where plant seeds dated to 14,400 years ago were found in desiccated human faecal matter. Dependence on plant foods also diverges from the Clovis model of a hunting-based society. Another site, dated to as much as 15,000 years ago, near Buttermilk Creek, Texas, yielded artefacts that may be precursors of Clovis technology, suggesting the origin of that culture from within the Americas rather than from Asia.

California's Channel Islands have several archaeological sites that support such challenges to the land bridge theory and provide evidence for migration by sea. Between 13,000 and 11,000 years ago, the Channel

Islands would have required a 9- or 10-km (about a 6-mi.) voyage from the mainland of North America. By that time, as the remains from several archaeological sites testify, seafaring peoples were living on Santa Rosa and San Miguel islands, which were probably first colonized 13,000 or more years ago. Significantly, no Clovis artefacts have been found at these sites, but projectile points similar to those found in the early layers of the Paisley Caves have. These people must have had watercraft sufficient to transport themselves from the mainland. Their descendants continuously occupied the northern Channel Islands from more than 10,000 years ago until the Island Chumash population was removed to the mainland in 1820, a story told in Scott O'Dell's novel, *Island of the Blue Dolphins* (1960). These people depended heavily on marine food sources for millennia.

Kelp forests along the North Pacific coast are among the most productive environments on Earth. At the end of the Pleistocene, they extended from Japan to Beringia and along much of the North and South American west coasts. These offshore, three-dimensional habitats foster varied and plentiful marine populations, including shellfish, seaweeds, fish, marine mammals and seabirds. Sea levels between 90 and 120 m (between 300 and 400 ft) lower than today's left vast tracts of flat, coastal land, now inundated, which by about 16,000 years ago provided an unobstructed migration route. Archaeologists dubbed this likely alternative to the land bridge the 'kelp highway', an avenue that would have allowed maritime hunter-gatherers to move speedily coastwise with plentiful availability of food and other material resources.

Coastal foragers enjoyed access to animals and plants that inhabited marshes, estuaries, rivers, coastal forests, sandy beaches, rocky shores – any of the varied ecosystems found in the borderlands between the ocean and the continental interior. The wealth of the littoral included seaweed; shellfish exposed at low tide; seals and other marine mammals that were vulnerable to hunters on land; the occasional beached whale; marine fish; salmon and other anadromous fish that returned from the sea to spawn in rivers; migrating birds; and also the full suite of land animals and plants. Initially coastal migrants may have used shores lightly, moving on when they exhausted a particular location, rarely settling for long periods of time. Reliable access to shellfish and other coastal resources, however, seems to have encouraged sedentary communities.

Habitable coastal areas from this time are now underwater, following the episode of rapid sea-level rise that ended about 7,000 years ago. We know that agriculture developed about 10,000 years ago. It appears increasingly likely that long-held assumptions about the path to civilization beginning with hunter-gatherers and proceeding through agriculture might be premature. Our understanding of the development of agriculture and civilization has emerged from the archaeology of interior sites, but new research suggests that coastal environments also supported large, sedentary communities and fostered the activities associated with complex societies and civilization. Sufficient food resources fuelled population growth in permanent coastal settlements and may have provided inhabitants with the time to produce craft goods, build public or even monumental structures, and experiment with game-keeping, gardening and active resource management. We can surmise that sea-level rise after the Last Glacial Maximum drove coastal people inland, perhaps promoting terrestrial lifeways. There is ample evidence, however, that the story of people and ocean is not only a long one but one in which connections between them were pivotal for the plot.

PEOPLE OVER THE millennia have used the ocean for many reasons, starting with food and transportation. The ocean, including its tides, currents, storms, exploitable resources, sea-level rises and falls, and other features, profoundly shaped human development from prehistoric times to the present. The impact is mutual and moves in both directions. The long story of people and oceans would not be complete without acknowledging that people have joined other elements of nature that act as agents of change in the ocean.

Humans made significant inroads into marine populations starting hundreds, if not thousands, of years ago. The number of Caribbean sea turtles is now in the tens of thousands, whereas a few centuries ago there were tens of millions. In studies going back 7,000 years, Caribbean island people targeting top predators including large fish and turtles caused local extirpations. Coastal inhabitants in what is now Maine likewise appear to have overfished cod locally, shifting fishing efforts to flounder and other species when cod catches declined. Evidence from middens suggests that Island Chumash people ate many large red abalone for a period of time between about 7,500 and 3,000 years ago, possibly because their hunting for sea otters removed the primary predator for abalone, enabling their numbers to increase. Recent partnerships between ecologists, historians and archaeologists have led to the conclusion that human exploitation, even long ago, could affect the size of fish available to catch, the geographic range of marine mammals or even the local marine ecology.

While the human relationship with the ocean includes cultural dimensions, explored throughout the rest of this book, that connection is firmly rooted in physical and ecological interactions. People, as part of nature, are inextricably connected to the ever-changing ocean. Their activities have, in turn, affected the ocean for millennia. Understanding this deep history makes apparent that the human relationship with the ocean started right at the beginning. The seas have provided food and transportation as long as people have lived near them, however long that may have been. Modern humans evolved eating seafood and living near coasts. As they migrated around the planet, groups of people diversified culturally, politically and economically in their uses of the sea and their understanding of its role in their lives.

TWO

Imagined Oceans

It is not possible to measure the full extent of the ocean
except with the eye of fantasy. No one will ever delve to
the bottom of that sea except by plunging into the waves
of his wildest dreams.

– From an account of a voyage to Siam (Thailand)
by a seventeenth-century Persian traveller[1]

STARTING IN PREHISTORIC times, communities whose subsistence or identities were inextricably tied to the sea had temporal and spatial connections to all three of its dimensions. Many coastal and island cultures employed their intimate knowledge of the sea to exploit its connectivity and its living and non-living resources, illustrated dramatically by such examples as long-distance navigation in Oceania, Ama diving and the practice of cormorant fishing. Thalassocracies such as the Phoenicians and Vikings projected power over targeted areas and fostered cultural identities that were tightly tied to the sea. The land-based powers around the Indian Ocean relied on navigation and trade as the bulwark of their terrestrial empires, while the spectacular voyages dispatched by the Chinese Ming dynasty during the first half of the fifteenth century subordinated the projection of naval power to the display of overwhelming wealth, with the aim of establishing tribute relationships. Chinese oceanic expansion, dramatic as it was, remained in the realm of the known world. Through the fifteenth century, people, commodities and ideas flowed easily and far, but oceans and seas were known and experienced individually, or as adjacent to a neighbouring sea, rather than as interconnected parts of a global ocean. Cultures around the world developed unique relationships with the sea reflecting the resources available to them, the geographic challenges and opportunities they faced, and also less tangible elements associated with their histories, spiritual beliefs and collective experiences.

MOST HISTORIES BEGIN when written records of the past are available. Since the human relationship with the ocean stretches back to evolutionary time, archaeology and folklore must inform ocean history. Prehistory has laboured under the same terrestrial bias as history, no doubt in part because inundation of coasts around the world following the end of the last great ice age has hidden sites of coastal settlement out of sight and consequently out of mind. If we embrace the broad definition of 'civilizations' proposed by the historian Felipe Fernández-Armesto as societies engaging in systematic refashioning of nature, then seaboard settlements should be added to the riverine, inland and desert regions where civilizations have been defined by agriculture, cities, writing or other traditional civilizing markers.

We know nothing of the first watercraft that carried humans, and even earlier hominids, to places we can demonstrate that they migrated to. Ancient cultural perception of the ocean likewise remains unknown, but perhaps not entirely unknowable. Examining how people used the ocean illuminates how they imagined the sea in relation to themselves. Knowledge of the ocean was derived from experience with the sea, such as long-distance voyaging, the familiarity with offshore banks gained by fishers or the ability to build and operate vessels to project power. Use of the sea promoted relationships between groups of people along coasts and across seas, establishing patterns of movement of people, goods and ideas that affected not only coastal communities but at times those far inland as well.

Ancient coasting enabled people to spread throughout the world and likely fostered settlement of coastal zones harbouring resources of both land and sea. The resulting communities and cultures diversified in response to geography, accident, historical experience and other influences, but they often shared features that may reflect common connections to the ocean. The populations of most seaboard civilizations consumed significant amounts of seafood, including finfish, shellfish and marine mammals, as evidenced by analysis of skeletal remains and also by the enormous shell middens associated with some of these settlements. The same types of knots have been found in northern Europe, Africa, Peru and Oceania, suggesting multiple, independent innovations taking place in locations once viewed as marginal to the story of human history.

The myths, folklore and superstitions of coastal cultures suggest extremely long-standing and important connections between people and the sea. The place of water in riverine and desert civilizations attracts more attention from folklorists than the mythology of coastal communities, but many of the world's oldest myths associate creation with water that is oceanic in scale. These oceanic origin stories are told in India, Mesopotamia and ancient Egypt, as well as in Mayan, Hebrew and Christian cultures. Many Native American peoples posited an initial primordial ocean from which an inhabitable world emerged, a sequence shared with other cultures north of the equator as far from the Americas as Siberia. Our understanding of the deep past leans heavily on examples of cultures that emerged inland and along rivers, but coastal hunter-gatherers spread to every part of the globe in the wake of the last great ice age, no doubt embracing oceanic waters as part of their myths and origin stories.

Mythology is often tightly connected to knowledge of the ocean. Coastal peoples of the Canadian Northwest hand down legends of escapes by canoe from inundations by the sea, and the survival of the Moken people in Thailand who, anticipating the 2004 tsunami, moved to high ground seems a modern instance of ancient knowledge of the ocean. Homer's evocation of Odysseus' choice to sail close to either Scylla or Charybdis, in myth a six-headed monster and a whirlpool, respectively, likewise reflects familiarity with the sea. As hazards on either side of the narrow Strait of Messina, influenced by strong tidal currents, the rock shoal and the natural whirlpool are commemorated in the sailor's expression, 'between a rock and a hard place'.

People spread out along the North Atlantic coasts 10,000 years ago. Europe has the largest ratio of shoreline to inland area of almost any part of the world. Its Atlantic coastline has greater biodiversity in its coastal regions and islands than the long-settled Mediterranean shores. The forests and meadows of the area called Doggerland between Europe and the British Isles provided resource-rich living space until sea-level rise about 8,500 years ago drowned those settlements. Although Europe's many peninsulas and wide coasts shrank, the cultures that remained facing the Atlantic shared the experience of depending on the rich and accessible resources of the littoral and of grappling with the ocean as a natural force challenging their efforts to use its restless waters to travel and move goods.

The common experience of plying the sea fostered social relationships that began with ceremonial gift exchanges and developed into trade. From Scandinavia to Brittany, to Iberia, ancient Atlantic communities built megalithic tombs for the collective burial of ancestors on islands or coasts that involved and relied upon the manipulation of giant stones, astronomical knowledge that might relate to ocean travel and, at times, burial in shell middens. As archaeologist Barry Cunliffe argues, peoples who lived along the Atlantic coasts of Europe shared common beliefs, lifeways and values over thousands of years, 'conditioned largely by their unique habitat on the edge of the continent facing the ocean'.[2]

The sea's threshold, the site of arrivals and departures, was faced with dread but also with expectation, viewed as a mysterious and supernatural place. The dangers associated with crossing liminal zones and going to sea have long been managed through ceremonies to propitiate gods before departing from land. Common practices and stories, such as of lands that sink into the sea, are known along and around North Atlantic coasts. Coastal communities there shared a firm belief, likely dating back to ancient times, that a drowning man belonged to the sea and should not be rescued lest the sea then claim the rescuer in his place.

Recent research at carefully chosen archaeological sites around the Atlantic, but also along the coasts of South America, the northern Pacific and other places around the globe, suggests that coastal societies relying on the resources of land and sea achieved population densities that stimulated cultural development. In such seaboard civilizations, fishing as well as trade, conducted initially by coasting rather than on long voyages crossing seas, invested a segment of the population with specialized knowledge and promoted the development of classes or social groups who took care of navigation or trade or warfare for their communities. Conceptions of ocean tenancy touched off conflicts with neighbours but intensive use of marine resources also motivated trade with inland or distant communities to exchange fish, shells or other coastal bounty for commodities unavailable to them. Access to plentiful food and other resources of the littoral zone appears to have fostered permanent communities that began to consider the coast and sea as part of their territory. In short, the sea has been central to the development of civilization. Although there were common features among ancient coastal communities around basins or even across the world, civilizations in different parts of the globe developed a diverse array of unique relationships with the sea.

THE FIRST BASIN TO BE spanned by human activities was the Indian Ocean. Its seaboard societies came to view the deep sea as a place outside of society. The ocean existed as a transport surface for trade but was understood as non-territory by regional powers that were more interested in spreading culture and promoting trade than projecting political power. This cultural understanding of the ocean emerged from the seasonal regime of monsoon winds and from the local coasting exchanges that over time developed into a trading network criss-crossing the entire Indian Ocean region. Starting probably 125,000 years ago – when *Homo sapiens* arrived on the shores of numerous seas that, with sea-level rise, merged into the Indian Ocean basin – people in this region fished and moved along coasts or crossed enclosed bodies of water in small vessels built from materials at hand: for example, dugouts, rafts or bark boats where wood was available, or reed boats in areas with marshlands. Navigation along the northern rim of the Indian Ocean may date to about 7000 BCE, initiated by networks of fisherfolk rather than by centralized or land-oriented powers.

The uneven distribution of both natural resources and manufactured goods motivated trade, initially over short distances but spreading to include wood and ivory of Africa, cotton textiles from India, spices from Southeast Asia and silk from China. The centralized state of Mesopotamia cultivated trade networks to exchange its agricultural surpluses for wood and stone from the western side of the Persian Gulf, but much trade remained small-scale and conducted by coastal fishing communities for thousands of years. By about 5000 BCE, long-distance trade connected Egypt, Arabia and the west coast of India. This trade involved communities in the littoral region; communities further inland were not as tightly connected to the emerging Indian Ocean world as their coastal counterparts.

In Hindu myth, life began in the sea or from primordial water, but when Vishnu is in the ocean, turmoil ensues. The cultural geographer Philip Steinberg argues that these beliefs inspired in some a fear of the sea, motivating mariners to cross it quickly and possibly stimulating innovation to do so. According to the Greek geographer Pliny, some Indian navigators carried birds to sea, releasing them if winds or currents pulled them away from land in hopes of following them back. Perhaps in this way sailors experienced first the waters just out of sight of land and eventually learned to cross bodies of water.

Characteristics of the Indian Ocean promoted the development of open-sea navigation, made possible by the evolution of larger vessels. In the western Indian Ocean, keel ships called dhows, built from planks laid side-by-side, sewn together with rope and then waterproofed with bitumen, became the workhorses of trade in the region. Clear tropical skies promoted observations of stars. The pattern of monsoon winds, once decoded around the first millennium BCE, enabled regular crossings from Arabia and western India towards Africa during the months of November to January, when the dry northeast monsoon winds carried vessels along. The opposing southwest monsoon, with accompanying heavy rains, enabled the return journey during the months of April to August. Surface currents complemented the monsoon winds, further accelerating ships travelling from north to south during the northeast monsoon and the opposite when the winds reversed. The same physical features that forged seasonal trade also worked to prevent its intensification. During the months of June and July, very strong winds often precluded sailing. The regime of monsoon winds stimulated seasonal trade to the east or the west but halted vessels for part of the year and prevented an increase in the number of annual transits.

The enforced idle periods for ships promoted the development of large and cosmopolitan port cities full of sailors, traders and travellers, a mostly male society that extended from the enclosed and separate world of ships onto land. Sailors on board vessels were generally subject to the laws and customs of the society from which they came. In ports, people from disparate parts of the region lived crowded together during the periods when travelling was impossible. Expatriate communities, given a significant degree of autonomy by host powers, controlled trade. Life in ports promoted the mixing of people and their cultures, spreading new religions, starting with Buddhism, Jainism, Hinduism and ultimately Islam.

The bubonic plague, which moved easily throughout this highly integrated region, disrupted trade in the sixth century. Slow initial recovery accelerated with the rise and spread of Islam and the near simultaneous establishment of the Chinese Sui dynasty (581 to 671), quickly followed by the Tang dynasty (618 to 907). Islam contributed a lingua franca in the Arabic language and also a set of widely acknowledged laws that benefited trade, while the consolidation of political power in China created a combination of political stability and demand for luxury goods that promoted economic exchange along what is known as the Silk Road.

Although goods, people, germs and ideas circulated throughout the region, waters were generally perceived as local or regional rather than oceanic. Only the waters off India itself were referred to as the 'Indian Ocean'. The waters to the west were the 'Erythraean Sea', a name initially applied only to the Red Sea and then extended outwards to include what is now the northwest Indian Ocean. The manuscript sailing directions known as the 'Periplus of the Erythraen Sea' documents the Indian Ocean trading system of 2,000 years ago and the entrance of the Roman Empire into that world. The Roman name *Mare prosodum*, or 'Green Sea', denoted what we now consider the central Indian Ocean at the latitude of Sri Lanka.

With the rise and spread of Islam, Arab knowledge of the Indian Ocean, based on the indigenous Indian local knowledge that solved the puzzle of the monsoon winds, increased and began to be systematically collected and recorded. Nautical charts featured carefully drawn coastlines showing landmarks important for coastwise sailing. The open ocean was not represented to scale but was, rather, left as simply abstract space to be crossed between sightings of land. In open water, Arabic navigators relied on stars and the knowledge, for example, of how high the pole star should be before turning towards their destination. To measure height, navigators employed a device called a *kamal*, made of wood and knotted string. They also used their knowledge of currents, winds, tides and the colour of the sea. By the time Ahmad ibn Majid wrote his late fifteenth-century treatise summarizing navigation in the Indian Ocean, Arabs had adopted the compass from China.

Arab and also Chinese knowledge of the world extended to the far corners of the Indian Ocean. An anonymous Egyptian authored the cosmological treatise *The Book of Curiosities* around 1020, which featured the first world map that included names of cities rather than only regions. The volume's Indian Ocean map used a Swahili name for the East African island of Zanzibar and depicted specific locations in China. A Ming dynasty world map created before the arrival of the Portuguese

Gravestone with a lateen-rigged sail in a 12th- or 13th-century Muslim cemetery in Istanbul, Turkey.

in the Indian Ocean likewise included accurate representations of parts of Southeast Asia, India and Africa.

Emerging land-based states such as in India and China had an interest in promoting trade and suppressing piracy, but kept ports and coasts insulated from interior regions. People inland were not entirely disconnected from the sea, as demonstrated by a sixteenth-century story of a guru worrying about a merchant captain devotee whose ship was stuck in the doldrums. Another devotee, fanning the guru and sensing his concern, blows the vessel to safety.[3] To most non-mariners, the ocean delivered desirable commodities but served mainly as a transport surface, a space outside society, but one where freedom of navigation and commerce was generally recognized.

Even after Islam dominated the Indian Ocean, its land-based states remained more intent on promoting trade and spreading culture than on projecting naval power or restricting navigation by other powers. On the eve of European arrival, this multi-ethnic region featured an integrated and extensive commercial system. For most people outside of ports and coastal communities, the ocean itself exerted a relatively minor presence in their daily lives. Sailors were likewise marginal, respected for their expertise in navigation and valued for their role in delivering goods but not considered part of society. The sea existed externally to society, constructed as a space devoted to trade. It represented distance to be crossed rather than territory belonging to the state.

TRADING VESSELS CONNECTED the Mediterranean Sea to the Indian Ocean thousands of years ago. In contrast to Indian Ocean states which did not view the sea as territory, ancient Greeks and Romans extended imperial domination seawards to a limited extent. The outer boundary of Greek geography was the forbidding Ocean River, flowing around the habitable world. Ruled by the deity Oceanus, the waters of the Ocean River were held as the font of all terrestrial rivers and streams, but Oceanus itself was mysterious, infinite and forbidding. The circle inscribed by Oceanus put the Mediterranean Sea in the centre of Greek geography. In Aesop's fables and in Roman mosaics, Thalassa, the primeval spirit of the sea, rises from the sea's surface holding an oar. Use of the Mediterranean for trade, fishing and the extension of power created a distinctly different human relationship to the sea than that forged around the Indian Ocean.

Physical geography laid the foundations for Mediterranean populations' use and conception of the sea. Due to flooding of coastal plains in the wake of sea-level rise, the Mediterranean coasts are narrower than those of either the Indian or the Pacific Ocean. Coastal dwellers, deprived of the resources of wide littoral plains, found themselves strongly oriented to the sea and detached from inland society. Marine fish provided an important source of food that was used first for subsistence but came to support commercial activity, while the uneven distribution of resources motivated coastal communities to trade with one another.

Before the Greeks, the mysterious Phoenicians, a confederation of maritime traders, occupied the coasts along the eastern Mediterranean. These two powers were the first to forge sea-based colonial empires. By the eighth century BCE, Phoenician trading posts ringed the entire Mediterranean. A painting in a rock shelter overlooking the Strait of Gibraltar depicting stick figures around a sailing ship may represent contact between Phoenicians and local people in the western Mediterranean who were unfamiliar with Phoenician maritime technology. The creation of the Atlantic port of Gadir (now Cadiz) enabled the Phoenicians to link the Mediterranean with the Atlantic for trade and other kinds of exchange. Gadir also became the centre for the earliest commercial sea fishery, which trapped the majestic bluefin tuna as it migrated through the Strait of Gibraltar returning to spawn.

Phoenician innovations in maritime technology enabled their domination of trade between eastern cultures in Asia and Iberia in the west. Thanks to mortise-and-tenon joints, Phoenician vessels were stronger than earlier designs that employed planks sewn together.

Phoenician ship carved on a sarcophagus, from the 2nd century CE.

Maritime archaeology has also revealed a lead-filled wooden anchor as a Phoenician invention. Ships and knowledge of the sea enabled Phoenicians to accumulate wealth and power, creating one of the earliest thalassocracies, a term describing seaborne empires or states with a significant maritime realm. Like that of the Greeks, the Norse and other similar powers that followed, Phoenician power derived from the sea rather than land.

Little is known about the Phoenicians, including what they called themselves: possibly Canaanites. Greeks coined the name 'Phoenician', perhaps referring to the prized purple dye produced from a type of mollusc. Phoenicians acted as middlemen not only for trade but for culture. They spread the modern alphabet, their creation, as well as myths and knowledge from Assyria and Babylonia throughout the Mediterranean, helping to set in motion the Greek cultural revival known as its Golden Age.

Throughout the periods dominated by ancient Greek and Roman cultures, the Mediterranean Sea remained in the centre of a widening world. According to the Greek historian Herodotus, Phoenician sailors circled around the African continent in the seventh century BCE. Yet when Homer composed the *Odyssey*, probably near the end of the eighth century BCE, the western Mediterranean was a relatively blank space to his culture. Historian John Gillis argues that gaps in Greek geographic knowledge abetted Homer's creation of the marvellous imaginative geography of his epic voyage tale. The similarly fantastic quest of Jason, who led the Argonauts to secure the Golden Fleece, contributed vocabulary to describe adventurers on the sea, and later in space. 'Argonaut' literally meant a sailor on the vessel *Argo*, but the word was adopted as the scientific name for the open-ocean mollusc, also called paper nautilus, and as a poetic term for mariners or adventurers, later inspiring the neologisms aquanaut and astronaut. By the time of Ptolemy, Greek geography had stretched to include China in the east and the Atlantic Ocean in the west. The Pillars of Hercules, the name given to the Strait of Gibraltar, separated the familiar Mediterranean from the chaos of the ocean beyond.

Trade was foundational for Greeks and Romans. Following the ancient Greek Rhodian code, Romans encouraged free trade, but this did not extend to a concept of absolute freedom of the seas. Rampant piracy prompted these coastal societies to project power seawards, while the region's geography, with the Mediterranean's many basins, gulfs and narrow straits, motivated efforts to assert state control to protect maritime commerce promoted by the long shoreline. Commercial fisheries, which by the second and third centuries employed nets, baskets and fish hooks, benefited from protection from piracy, particularly as their products were drawn into long-distance trade. Production and trade of salted fish and fish sauces spread Roman tastes throughout the empire. Salt-cured tuna packed in amphorae travelled hundreds of kilometres, perhaps more. So important was the tuna fishery that the citizens of several cities that exploited this resource used images of tuna on their coins. Fish and fishing also gained symbolic meaning for early Christians, who employed the fish symbol to signify their community before they could worship openly.

The Greeks and Romans considered the sea, which they dubbed *mare nostrum* or 'our sea', central to their world, but they never embraced the ocean as a natural element for people. Navigators preferred coasting to sailing out of sight of land across wide stretches of water. Sea battles took place close to land, not far out at sea. Sailors as a rule did not sleep or eat meals on board vessels if they had the option of going ashore. Although they preferred to stay close to shore, these ancient military powers extended operations under the sea, deploying divers to cut anchor cables, sink ships and construct harbour defences. Mediterranean powers did not view the sea as a space entirely outside of society or immune from state power. The Roman Empire asserted the right to extend its power to protect sea trade, tie peripheral areas to its centre and assure its interest in marine resources, including the resource of connectivity. The sea was understood not as land-like territory that could be possessed but as a sphere of influence where sufficiently powerful states could intervene. As a consequence of their regional hegemony, Romans assumed guardianship over coasts and enclosed waters but eschewed the open sea.

THE COLLAPSE OF THE ROMAN EMPIRE coincided with an inland turn as feudal Europe grew strongly agricultural and abandoned ocean fisheries in favour of freshwater fishing. Scandinavia and other northern areas continued to employ marine food resources for survival and maintained a culture of seagoing, shipbuilding and strong orientation towards the ocean. Despite the generally terrestrial focus, islands, promontories and coastal peninsulas in Christian northern Europe retained special significance tied to ancient practices and mythology, often serving as the location for burial grounds, holy places or forts.

Some holy men turned deliberately to the empty, desolate ocean of Europe's so-called Dark Ages. Viewing the sea as akin to the desert, penitents and monks set sail, particularly from Ireland, into the open Atlantic Ocean in the fifth and sixth centuries. Voyages were spiritual quests rather than destination-based travel, although hermits sometimes managed to survive on rocky outcrops of remote coasts or isolated islands where they landed. Such groups of holy men established monasteries in places as far away as the Faroe Islands and even Iceland.

Norse raiders expelled monks and hermits from their isolated coastal retreats throughout the North Atlantic. The seagoing prowess of the Vikings was unparalleled throughout Europe. Until the eighth century, they were known for trading and raiding, activities that accrued wealth and status. More amphibious than truly oceanic, Vikings were as tightly tied to land as to the sea, conducting raids during breaks in the agricultural cycle of work. They understood boats as transportation not just for the living but to carry the dead to the next world, burying them in graves inside ships. By the early ninth century, a combination of shipbuilding innovations produced the classic longboats, powered by square sails in addition to oars, which enabled fast and efficient travel across the oceans.

Vikings moved in all four cardinal directions along North Atlantic coasts and the river systems of Europe. Population pressures and social dislocations prompted migration and new settlements from the Mediterranean north to Iceland and from the Caspian Sea west to Newfoundland. Vikings undertook their long, ambitious voyages carrying seeds and animals with which to recreate their familiar livelihood and grow their communities in new lands. Norse mariners ventured north of the Shetland Islands perhaps with knowledge of the earlier successful voyages of Irish monks, reaching the Faroe Islands in about 800, Iceland less than sixty or seventy years later and Greenland within a few decades after that.

While history considers the Vikings to have set forth boldly to cross the open ocean, reaching Newfoundland on the far side of the North Atlantic by around 1000, these intrepid voyagers understood their own travel differently. They moved in manageable steps around an ocean space they conceived as an enclosed sea. Their imagined geography was bounded by the Norwegian shore, Greenland, Baffin Island, Newfoundland and Africa. Rather than discovering a new world, Leif Erikson coasted along an inland sea, coming across the next auspicious shore for settlement. Such mental enclosure rendered the North Atlantic less terrifying, although it certainly posed a formidable barrier that required skilled navigation, seaworthy vessels and a maritime-oriented culture.

Unlike the Vikings, most European cultures faced away from the sea during the medieval period. While the Vikings and other seagoing peoples helped tie Scandinavia to Europe and integrate western and eastern Europe, their trade focused on prestige and luxury items and some specialized goods. Europe turned back to the sea partly through a rise in international trade fuelled by a commercial revolution but also out of necessity. Depletion of freshwater fisheries, such as sturgeon, salmon, whitefish, shad, eel, pike, bream and trout, prompted a rather sudden revival of sea fishing in the eleventh century. Evidence for this swing comes from analysis of kitchen middens, which reveal an abrupt shift from 80 per cent of consumption of freshwater species to 80 per cent of marine species, especially haddock, cod and herring.

Woodcut depicting herring fishing in Scania, Sweden's southernmost part, 1555.

Commercial fisheries prosecuted by specialized workers burgeoned from this time, contributing marine products to expanding trade routes and markets. The Basques conducted the first commercial whaling in the Bay of Biscay in the High Middle Ages. They may have moved to the western North Atlantic even before Columbus's voyage, in pursuit of Atlantic right whales, whose population they decimated in their own nearshore waters. Markets for oil, fur and ivory sent Europeans searching the North Atlantic for the hitherto mysterious creatures that yielded these valuable commodities. Whereas Vikings sailed across what they perceived as an enclosed sea seeking trading partners or sites for settlement, and Irish monks set sail to seek holiness or penitence, Europeans in pursuit of whales, seals and walrus may have been the first to explore the sea itself.

By the eleventh century, networks of trade bound Europe together as markets developed to serve the needs of emerging states. Medieval mariners had mastered the Strait of Gibraltar, but regular traffic awaited commercial development that flowed after Crusaders returned with a taste for goods from the East. Trading centres grew up at coastal ports facilitating exchange of food, raw materials and imported goods. Trade eventually extended from Muslim Iberia to Viking colonies in Ireland and even Iceland. Between the Frankish and Byzantine empires stood Venice, the city-state that served as a fulcrum for trade between east and west. Like the Phoenicians, Greeks and Vikings, Venice and other Italian port cities were land-based, coastal maritime powers that projected military power over the sea in order to dominate trade, including controlling access to seas and routes, but did not consider the ocean itself claimable space.

The Swedes, who controlled the Baltic in the twelfth and thirteenth centuries, also ruled by projecting power over the sea, accumulating great wealth in the process. But like other Baltic and North Sea powers, including the famous Hanseatic League, Sweden exercised an element of control over waters to be deemed its own. This northern European posture towards ocean space was fostered by differences in the marine

Inuit driftwood map, read by touch to navigate along the coastline.

environment relative to southern Europe, whose mariners enjoyed clear waters and almost constant daytime visibility of coasts instead of fog and storms that required navigators to employ sounding devices, charts and specific knowledge of particular ocean sites. Prolific fishing grounds in the north also encouraged households, feudal estates and states to think of adjacent waters with rich fisheries as annexes of terrestrial claims.

LIKE THE ATLANTIC, the Pacific also hosted cultures whose heavy use of marine resources translated into a sense of ownership of the sea. Recent archaeological research suggests that marine hunter-gatherers in coastal areas of present-day Peru, the western United States and British Columbia achieved high population densities and were characterized by rich and socially stratified cultures long before agriculture appeared. While the Aleutian Islands have limited terrestrial resources, marine and coastal resources supported large, stable villages that fostered complex political and social life among the Aleuts. From other northern regions, Inuit maps survive to reveal the practice of making usable representations of the coast: pieces of driftwood carved to depict a section of coastline that were small enough to fit inside a mitten, which the navigator could read in the dark by touch.

The Pacific Northwest coast remained off Western world maps longer than any other inhabited continental coasts. Such groups as the Makah of Washington State understood marine space and its resources, including fish, whales and seals, as belonging to them. The kind of ownership long understood for land and terrestrial resources extended to the sea for many Pacific cultures, both coastal and island. The Makah applied their intimate knowledge of the sea to use its bounty, at the same time respecting sea creatures as part of the spiritual world. The revival of whaling by the Makah in the twentieth century represents, in part, a reassertion of their claim to the sea.

In contrast to the Atlantic and Indian Oceans, nowhere on our planet have people seemed to be more comfortable with the ocean than in the Pacific. This gigantic ocean is twice the size of the Atlantic and greater than all of the Earth's landmasses. It is the globe's largest natural feature, touched by five continents. It is a space of superlatives, both in its physical and cultural geography. Its more than 25,000 islands are home to people who collectively speak a thousand languages. In settling these many islands, Pacific people set sail well before recorded history and learned to voyage long distances to places far out of sight.

Three regions are generally recognized within the Pacific: Melanesia, Micronesia and Polynesia. Melanesia, stretching from New Guinea to Fiji and lying north and east of Australia, was populated first. Micronesia, to the north, and Polynesia, to the east, were settled later. Some observers criticize the identification of the so-called three 'nesias' for oversimplifying reality, yet Pacific peoples have themselves adopted these categories for their utility for certain social or historical purposes. Originally, the Pacific was understood by its inhabitants not as three large regions but as a series of seas. Our present conception of the Pacific as a unified ocean dates from European efforts to comprehend these waters in the fifteenth through eighteenth centuries.

The peopling of Pacific islands began between 40,000 and 50,000 years ago, when lower sea levels rendered the current New Guinea and Australia into the single continent of Sahul. On the eastern end of New Guinea, the Bismarck Archipelago extends towards the Solomon Islands. This area has been called a 'voyaging nursery', where people learned to cross first short, and then longer, stretches of water as they gained experience and competence at sea.

More details emerge for the period around 3,500 to 4,000 years ago, with the arrival in Melanesia of a new cultural complex named 'Lapita' by archaeologists and characterized by its distinctive pottery. Lapita people moved pottery and other objects as well as natural resources such as pigs, chickens and dogs as they migrated and settled throughout Melanesia. From DNA evidence, we know that these people also moved from Taiwan southwards to the Philippines before entering Melanesia. For about 30,000 years, the eastward limit of human settlement in the

Pacific, outside the Australian landmass, had been the Solomon Islands. The Lapita peoples moved out of the Solomons by 3,200 years ago, reaching Fiji at the easternmost tip of Melanesia by around 3,000 years ago. The voyage to Fiji required crossing 850 km (528 mi.) of open ocean.

Micronesia and Polynesia were settled next, more or less simultaneously. Within a few centuries of reaching Fiji, sites appeared on Polynesian islands including Samoa and Tonga. Micronesia was likely settled first by people who shared a common ancestry with the Lapita but arrived directly from Taiwan or the Philippines. Second came Lapita migrants moving northwards from Melanesia. Micronesia's islands are smaller and farther apart than those in the Melanesian archipelago of Fiji or those of Samoa and Tonga to the east. Polynesia includes islands lying over 4,000 km (almost 2,500 mi.) apart, stretching over a gigantic triangle of several millions of square kilometres bounded in the corners by Easter Island to the east, Hawaii to the north and New Zealand to the south.

It is a striking achievement that members of a kindred group of humans found and settled all these islands in a relatively short period of time. Carving out a living on coral atolls required innovations to subsistence patterns and habits as well as to social organization and culture. Pacific peoples carried with them animals, and also plants such as taro, yams, breadfruit, bananas and sugar cane when they travelled to settle new islands. Double-hulled vessels up to 30 m (100 ft) long could carry as many as 250 people on short trips. For long voyages of migration, such a craft could accommodate perhaps 100 people and the tonnes of goods and living things needed to establish a comfortable life on a new island. Adaptation and innovation characterized settler communities, but many components of subsistence were spread widely through the Pacific, along with elements of sophisticated political and religious systems.

The ability to sustain communities on atolls did not rest exclusively on the terraforming activities of settlers. Life on Pacific islands equally drew upon fishing and aquaculture. The boundary zones between land

and sea, including tidal flats, estuaries, reefs and lagoons, provided plentiful and accessible resources. Fishing around reefs and constructing fish traps also drew upon nature's bounty, but islanders learned to raise fish in ponds where they were fed with cultivated algae or nutrients from seawater entering through sluice gates or other means.

Not every island proved able to sustain life. Some were never colonized, or at least no archaeological evidence exists of attempts to do so. That is why the mutineers from the crew of HMS *Bounty* were able to retreat to Pitcairn Island, uninhabited in 1789 when the infamous mutiny occurred. The enigmatic Rapa Nui, better known to Westerners as Easter Island, illustrates the ecological limitations of islands. Probably settled early in the second millennium CE, Rapa Nui sustained a thriving culture whose people created and installed monumental statues between the twelfth and seventeenth centuries. Called *moai*, these enormous stones posed a mystery to the European explorers who arrived in 1722 to find a deforested island. Experts in the past believed that the island had hosted a population of 15,000 at its height, with humans bearing the responsibility for ecological disaster, but today some archaeologists think that 3,000 may have been the maximum population and that Polynesian rats gnawing palm seeds contributed significantly, along with human activity, to denuding the island of trees.

Rapa Nui and Pitcairn are among the most remote islands on the planet. In contrast to residents of these exceptional islands, many Pacific islanders maintained firm connections to communities on nearby – or sometimes quite distant – islands, participating in trade and cultural exchanges. A well-articulated example is the Kula ring, an exchange system in the Trobriand Islands to the east of New Guinea. Necklaces with red shell discs are traded in a clockwise and northern direction in ceremonial exchange for white shell armbands that are passed along southwards. Evidence from oral history on the islands of Yap and Palau in the Caroline Islands extends this trading network back at least ten generations, but some scholars believe it may date from the fourteenth or possibly even the twelfth century. In his 1922 groundbreaking study,

The Argonauts of the Western Pacific, Bronislaw Malinowski explained the Kula system of reciprocity and exchange linked to political authority. His work established the importance of fieldwork for anthropology and contributed to an understanding of Pacific culture as occupying regional seas and extending across and connecting social and political zones.

More recently, the anthropologist and writer Epeli Hau'ofa reframed the perception of the Pacific world created by Europeans, attempting to recover an original, and continuing, islanders' worldview. Instead of small, isolated islands almost lost in the enormity of an ocean, Pacific peoples experience 'seas of islands', to invoke the title of Hau'ofa's influential 1993 essay. The decentring of land melds with the cosmology of the people of Oceania, for whom land surfaces represent, in myth, legend and oral history, a minor proportion of meaningful territory relative to the ocean, the underworld and the heavens. Oceania is a profoundly three-dimensional place, extending into the sea and skywards in physical space and correspondingly in the spiritual realm.

Hau'ofa's insight is not restricted to the past but has deep political reverberations in the present. He identified as a product of imperialism the European construction of Pacific islands as too small and resource-poor to support economic growth. This interpretation defines Pacific peoples as isolated and confined to their small island homelands. Hau'ofa instead argues that ordinary people of Oceania today are as unconfined by imperially etched boundaries as were their navigator ancestors. Today Pacific islanders travel globally between islands and mainlands, working and accumulating property, transporting and exchanging resources, and exercising and cultivating kinship networks. Construction materials or cars, for example, might be moved from homes abroad while island-made handcrafts, kava (a root found in the South Pacific), dried marine food sources or island agricultural produce might flow in reverse. 'So much of the welfare of ordinary people of Oceania depends on an informal movement along ancient routes,' declares Hau'ofa.[4]

Pacific peoples are highly mobile, in the present via air travel but historically because they continue to experience the sea as home. The geographer Philip Steinberg contrasted the understandings of the ocean in cultures around the Atlantic and Indian Oceans with those of the Micronesians. Unlike other peoples, for whom the sea was generally a space outside society, Micronesians considered ocean space akin to land space, rendering it subject to territorial control. The ocean space of one island adjoined that of another, with no unclaimed area between. Yet, the claim asserted was not one of abstract possession. Rather, people valued ocean territory for its resources, whether productive fishing grounds or the attribute of connectivity between islands for desired trade or social contacts. Fishing areas were abandoned if better ones were identified or if they stopped yielding resources. Access to transit was controlled through the management of navigational knowledge, particularly through the social organization of elite navigators.

The sea as a space of society and a place of movement is a feature of the histories of Pacific islanders. Most island histories begin with their own discovery and settlement, in a narrative that featured arrival from a distant location, often understood as coming from the west. Even in western Polynesia, where origin stories do not tend to have strong traditions of geographic source, people understand their ancestors to have arrived from far away. Polynesian navigational concepts and techniques match those of Micronesia, suggesting the importance of ocean voyaging for the dispersal of people throughout the Pacific.

Western anthropologists have long argued about whether the spread of people around and across the Pacific happened by accident or as a result of deliberate voyages. A pragmatic response to this debate might be to question the long-held assumption that these must be mutually exclusive possibilities. While accidental drift might potentially explain some migrations, scholarship based on computer modelling concludes that several critical stages of dispersal were not likely explainable other than by intentional sea travel. According to Thomas Gladwin, who in 1970 studied traditional navigation of the people of Puluwat Atoll in Micronesia, the year 1945 was the last instance a canoe was lost at sea. Sailor and student of traditional Polynesian navigation David

In 2017, the Polynesian double-hulled voyaging canoe *Hōkūle'a*, launched in 1975 to revive voyaging culture and knowledge, completed a three-year circumnavigation blending traditional and modern technologies to promote *Mālama Honua*, which translates to 'caring for our island Earth'.

Lewis cited this evidence in support of his contention that long-distance voyaging not only fostered connections between places but also enabled intentional migration and surely increased the likelihood of successful outcomes for accidental drifts.[5]

Traditional navigation in Oceania draws upon sets of concepts and techniques passed down orally through a lengthy apprenticeship involving memorization of the motions of stars, the weather at different seasons, and the habits of certain animals. Unlike in European navigation, instruments are not used. Once the direction for travel to a particular place is chosen, steering is guided by the setting point of a star or by keeping the vessel's track on a particular angle to the Sun, to the sea's swells or to the wind. Homing birds guide alert observers to islands out of sight, while changes to the set of the current or particular cloud formations or types of phosphorescent organisms might likewise signal the approach of an island. Navigation was not limited to the sea's surface and the heavens, but sometimes included gathering information from the ocean's third dimension, as when a navigator sat cross-legged or lay down on the bottom of his canoe to feel with his body the ocean swells formed by trade winds and far away storms that rolled underneath the currents at the surface.

Starting in the 1960s, scholar-sailors have tried to learn the secrets of Pacific navigation by studying with expert navigators, through oral history, and by conducting experimental voyages on traditional and replica vessels as well as modern ones. Historically, the knowledge of wayfinding at sea and the construction of ocean-going canoes was closely guarded and kept among a powerful and high-status group on each island. Modern investigators have sought out individuals who had conducted long-distance voyages or still did. Several found Hipour, a Micronesian master navigator from Puluwat, whose teaching of several Western scholars spurred interest in traditional Pacific navigation and promoted the resurgence of long-distance voyaging and navigational training in the Caroline Islands and elsewhere in Oceania. A Polynesian expert named Tevake could point accurately towards out-of-sight islands and make landfall after 72 km (45 mi.) of travel with no view of the sky.

Scholars conclude that master navigators of the Pacific did not, except during training, compartmentalize the tasks involved in their work. Rather than 'steering a course' or 'fixing a position', they integrated the collective input from stars, wave swell and marine animals to gain an understanding of their precise position, their trajectory and the best strategy to reach their destination. The navigator-priest Tupaia, who sailed on Captain James Cook's expedition to New Zealand and Australia from 1769 to 1770, could always point during the voyage to the direction of his home island of Tahiti and knew all the Pacific island groups except Hawaii and New Zealand. The art of navigation was not only rooted in the social structure of island communities but, as Tupaia's status suggests, had ties to spiritual life and belief, rendering navigation as an expression of the tight connections between Pacific peoples and the ocean, associations that were as cultural as they were physical.

Pacific islanders gazing out at their many 'seas of islands' did not, however, have a word or concept for their vast ocean in its entirety. The enormity of the Pacific was perhaps mitigated for them by the sheer number of islands that filled their cultural perception of their watery world. By contrast, Europeans, who first perceived of the Pacific as a whole, came to view it as a void, possibly because it took them over four centuries to find all the islands long known and inhabited by the peoples of Oceania.

WHILE ISLANDERS CONCEIVED of their worlds as seas of islands, many people living around the rim of the great ocean experienced the Pacific as a series of connected seas. This was true of inhabitants of Japan, which has, at times, been viewed as part of the Pacific and at other times not. Since the Palaeolithic period, people in Japan have relied heavily on the sea, particularly because its wide mountainous regions and limited freshwaters restricted agriculture and inland fishing. Shellfish, seaweed and prolific salmon runs provided sustenance, while meat and other parts of stranded whales were considered gifts from heaven. Hollowed-out boats enabled trade networks for rare stones and exotic shells, and also provided coastal transportation. Seasonal migration maximized food sources available for subsistence. Later, bone and horn harpoons and hooks and nets with clay weights enabled the exploitation of fish, marine mammals, shellfish and seaweed. Early fishing culture resembled that of the Eurasian subarctic, while later innovations appear to have arrived from Malaysia and China. As in China, fish were long considered a symbol of luck and prosperity. Fishing and other uses of oceanic resources remained important to Japanese culture and economy even after a more agrarian society based on rice cultivation began to grow about 1,000 years ago.

As in many places, the story of tight connections between Japan and the sea stretches far back in time. These island people also knew the ocean intimately, as several distinctive practices attest. Both fishing

Katsukawa Shunsen and Kawaguchiya Uhei, colour woodcut of women cormorant fishing at night, 1800–1810. One is at the helm and one is breastfeeding, both watching the third holding a torch and the leashes of the three cormorants.

with cormorants and the remarkable free diving of the Ama extended human links to the sea's third dimension. As early as the seventh century, cormorants were employed to help people catch fish. This activity drew upon the knowledge that cormorants dive to pursue their prey and can propel themselves underwater, to depths of up to 45 m (148 ft), with their webbed feet and their wings. Fishing with these birds seems to have developed independently in Peru and also in the Mediterranean, and is still practised on the Nagara River in Japan today. The fishing master controls the bird with ropes, and a bird can capture but not swallow the fish because of a ring affixed around its neck. Originally a means to catch fish for subsistence, cormorant fishing in Japan became an imperially sanctioned activity whose first annual catches are dispatched to the capital.

The roots of the famous Ama divers date from about 2,000 years ago in villages along the Japanese coast where breath-holding divers fished for food to supplement harvests from land. Initially both men and women pursued shellfish, seaweed, octopus, sea urchins and fish.

Later the practice of diving to collect shellfish and seaweed transformed into a women's occupation. These divers developed breath-holding techniques and diving practices that enabled dives as long as two minutes, reaching up to 30 m (100 ft), although mainly concentrating in depths of about 9 m (30 ft). The fishery shifted from its subsistence orientation to a focus on abalone at some point between the eighth and twelfth centuries. Abalone, found in prehistoric mounds around Japan's shores, long held religious significance as an offering to deities and gained cultural importance as a luxury commodity. Dried abalone became a major export prized in China from the seventeenth century on. In the late nineteenth century, the creation of the cultured pearl industry tended by Ama divers included their performance for an emerging tourism sector.

Many Japanese maritime traditions that leaned on intimate knowledge of the marine environment, including fishing with cormorants, spread from China. There, dense populations had long depended heavily on freshwater fishing and, by about 1,400 years ago, fish culture as well. Though China boasted far more inland territory than Japan, its coastal areas were from the beginning oriented towards maritime activities, even as the colossal empire concentrated on terrestrial development. In this way, China resembled Indian Ocean societies, with land-oriented leadership existing alongside an ocean-oriented zone along the coast. Coastal fishing, though practised far back in time, was at times inhibited by piracy or typhoons. While dugout canoes, so prevalent in other parts of the Pacific, did not develop in China, seaworthy junks, likely derived from rafts used on rivers, appeared before the year 1000.

Chinese merchants were involved in Indian Ocean trade by the fifth or sixth century through Canton and other entrepôts, but the government did not establish a navy or project power overseas until the twelfth century. With the compass, a Chinese invention of 200 BCE but first used for navigation at sea after 1000 CE, and with other navigational instruments and techniques borrowed from Arab and Persian navigators, Chinese vessels set forth to establish tributary relationships with

Replica of a medium-sized treasure ship (63.25 m/207½ ft long), built in 2005 from concrete and wood, commemorating Zheng He's fleet and located at the Treasure Boat Shipyard, Nanjing, China.

neighbouring areas. Trade remained important, and Chinese tastes and market demands emerged to exert an outsized effect on the increasingly linked economies around the Pacific.

During the first third of the fifteenth century, China sponsored a series of major voyages that reached Southeast Asia and India, and eventually the Persian Gulf and Africa. The emperor who instigated these ambitious voyages, known as the Yongle Emperor, added the ocean trajectories to his inland efforts to expand the Grand Canal and the Great Wall and to demonstrate the Ming dynasty's power against the Mongols. At this time, China had been active in trading by sea for several centuries and led the world in naval technology and shipbuilding. Innovations included the construction of vessels with watertight compartments and sternpost rudders, as well as the replacement of rowers with masts and efficient sails, freeing space for cargo. Already by the eleventh and twelfth centuries, in addition to adopting Arab navigational innovations, Chinese shipbuilders had borrowed lateen sails, permitting more effective upwind sailing. Printed manuals that included novel star charts and employed compass bearings provided a tool for navigation when skies were clear. Most of these developments were in use for about 1,000 years in China before appearing in Europe.

When the Emperor decided to reach outwards into the seas surrounding China, he selected a trusted and talented eunuch military officer and diplomat who had served his household since his capture as a boy from the Mongol province of Yunnan. Zheng He was appointed in the second year of the emperor's reign, in 1403, to oversee the construction of the fleet and lead the voyages. The first voyage sailed two years later, visiting Vietnam, Thailand and Java before reaching Calicut (Kozhikode). There the fleet remained for trade and diplomacy, but also to await the shift to the monsoon regime in spring in order to return in 1407 using favourable winds. A second voyage departed soon after the return of the first, repeating the journey to Calicut to attend the inauguration of a new ruler. Similarly, the third voyage set sail in 1409, the same year the second returned, this time adding Malacca in Malaysia and Ceylon (Sri Lanka) as ports of call. In 1413, two years after the third voyage returned, the fourth embarked with a more ambitious geographic goal. The main fleet reached Hormuz in the Persian Gulf, while detachments moved south along Africa's east coast, reaching almost to Mozambique.

All the voyages involved trade and the establishment of tributary relationships. More impressive even than the distances covered were the sizes of the fleets and vessels dispatched. The first voyage involved more than three hundred ships, sixty of which measured over 121 m (400 ft) in length and featured nine masts. These gargantuan treasure ships were meant to display the overwhelming technology and wealth of the Ming dynasty to foreign powers. Accompanied by hundreds of smaller vessels, the fleet transported cannon, horses, troops and water, as well as gifts such as silks, porcelains, tea and iron. Often diplomats from foreign powers accompanied the treasure ships back to the Chinese capital. Many such officials were returned on the fifth expedition, which reached Aden at the mouth of the Red Sea and revisited the east coast of Africa. The sixth voyage reached Mozambique, as far south as any of the Chinese fleets ever ventured.

The Yongle Emperor died two years after the sixth voyage returned. The seaward expansion, although popular with a pro-expansionist faction within the court, was opposed by conservative forces that objected to the extravagant costs. The son who inherited the throne cancelled the voyages but died shortly after, leaving as his successor a son who favoured his grandfather's policies and, as the Xuande Emperor, ordered a seventh voyage that got under way in 1430. Zheng He died during or shortly after the last leg of the voyage, before which he installed an engraved pillar recording for posterity the landfalls and achievements of the first six voyages. In the eclipse of ocean-oriented trade and exploration that followed, no further voyages were dispatched, ocean-going vessels were destroyed, and prohibitions against building ships with more than two masts were enacted. A navy that had boasted over 3,000 ships essentially ceased to exist as China turned decisively inland and mostly forgot these spectacular and intriguing voyages until the twentieth century.

ALTHOUGH HISTORIANS ARE not fond of counterfactual analysis, it is tempting to consider what might have happened had China continued its maritime exploration and Europeans met treasure ships and gigantic, well-equipped Chinese fleets. However, it is also important to recognize the distinct differences between Zheng He's voyages of the first third of the fifteenth century and subsequent European maritime expansion beginning with the Portuguese explorations of the same century. The impressive fleets were not exploring unknown territory, but following routes forged by Chinese traders. The purpose of the voyages was not primarily trade but instead to display China's power, ingenuity and wealth and to induce foreign powers to offer tribute to the Ming dynasty. The story of China's abandonment of ocean voyaging reveals a struggle over embracing an identity as an oceanic culture and eschewing the sea to pursue isolation and terrestrial priorities. As the Chinese case illustrates, knowledge and use of the ocean connects intimately to cultural intentions, choices and desires. Chinese ambitions drove Zheng He's fleet to pursue the knowledge and technology to accomplish its impressive voyages. On the other hand, that knowledge and technology was within reach due to earlier uses of the sea by Chinese traders, shipbuilders and others.

Humans have always been connected to the ocean, for subsistence and transportation to start. Different cultures evolved distinctive connections to the ocean shaped by geography but also by experience of the sea and use of its resources, involving knowledge of the ocean but equally imagination. Vikings understood their seafaring activities as confined to an inland sea, a vision that enabled them to cross the fearsome Atlantic. People of Oceania saw seas of islands where European explorers saw a vast and empty ocean. Communities around the world, including major land-oriented empires, exhibited tight interrelationships with the ocean that extend far back in time and profoundly shaped their respective and diverse cultures. Seagoing and coastal peoples through the fifteenth century forged linkages between each other that spanned regions, basins and even joined seas. The mariners who sailed after that time would find connections between all the Earth's oceans and seas, creating a global world through the same patterns that tied earlier peoples to the sea – namely, the enterprise of combining experience with imagination to know and use the ocean.

THREE

Seas Connect

The ebbs of tides and their mysterious flow,
We, as arts' elements, shall understand,
And as by line upon the ocean go,
Whose paths shall be familiar as the land.

– John Dryden, from *Annus Mirabilis* (1667)

COASTAL PEOPLES HAVE always lived by and with the sea, relying on its resources and using its waters for transport as well as cultural identity. Although the imperial Chinese fleets of the early fifteenth century reached East Africa and the Persian Gulf, people did not extend voyaging, or their geographical imaginations, to a global scale until the fifteenth and sixteenth centuries. Those who did so were Europeans, whose pursuit of wealth through trade and colonization expanded beyond the coasts and basins of their part of the world to encompass the Earth. Traditional understanding of this period viewed explorers' activities as motivated by the goal of finding new and unknown lands, or the source of valuable commodities such as much-desired spices from the east. In his landmark book *The Discovery of the Sea* (1974), the distinguished maritime historian J. H. Parry proposed that the so-called 'great age of discovery' yielded knowledge about sea routes. The originality and significance of explorers' achievements lay less in discovering new lands than in linking inhabited and known lands by ocean routes. The truly novel discoveries were not terrestrial but oceanic: the recognition that all the Earth's oceans were connected and the knowledge of how to move between all the oceans not covered with ice. With this framework in mind, Portuguese navigators discovered the sea route around Africa to the Indian Ocean and on to the Far East, while Columbus's immediately important discovery was the realistic possibility of a round-trip voyage across a bounded ocean basin. For many groups of people throughout history, such as Micronesians or Vikings, it would be meaningless to talk about a 'discovery' of the sea. For modern Western history, though, discovery of the sea constituted a profoundly consequential category, because knowledge of the ocean as a global feature underlay the extension of power and exercise of imperial control by European nations that created global economic, technological and cultural networks.

CLOSURE OF THE EASTERN Mediterranean to Europeans in the mid-fifteenth century, by the fall of Constantinople to the Turks, cut Europeans off from trade with the Far East and stimulated exploration. By this time, combinations of routes by sea and river connected all of Europe, facilitating the easy transport of trade goods, ideas, and also germs, including the bubonic plague that set off the Black Death, which killed a third of the human population of Europe. The expansive trading network forged by connecting southern with northern Europe created a vibrant maritime commerce that defined both Italian port cities and Hanseatic towns and their merchant communities. Disruption of the overland Silk Road as the Mongol Empire unravelled and plague spread, along with a rise in Mediterranean piracy, prompted European merchants with capital to spare to look for alternative trading routes.

Portuguese navigators began coasting voyages along the northwest African coast in 1419, under the direction of Henry the Navigator, soon also reaching the islands of Madeira and the Azores. By the mid-fifteenth century, highly manoeuvrable caravels had become the craft of choice for oceanic voyages because their seaworthiness and relatively shallow draft rendered them effective for operating in coastal waters. Although it is tempting to assume that Europe's successful search for new sea routes might be owed to technical or knowledge-based factors, recall that Zheng He's fleets had equivalent, if slightly different, vessels, rigs and navigational knowledge. So Europe's achievement rested on other circumstances. Within Portugal, these included a belligerent

A Portuguese caravel of the 1560s.

Christianity that, extending from the legacy of the Reconquista, sought to suppress expansionist Muslims; the very great attraction of trade in gold and slaves from Africa's coasts; a chronic shortage of arable land and foodstuffs; and ultimately the allure of spices from the Far East.

The discovery of the sea required mariners to abandon coasting for open-sea sailing. The Portuguese rounded Cape Bojador in 1434, surmounting a significant psychological obstacle that opened up exploration of the western African coast. The Gulf of Guinea, which became the headquarters of Portuguese slave trading, lay in the doldrums, where ships might drift for weeks, powerless against unfavourable currents. To the south, coastal prevailing winds and currents ran counter to the direction the explorers wanted to travel. Bolstered by their island discoveries, the bold Portuguese captains who mastered the route between Lisbon and the so-called Gold, Ivory and Slave Coasts applied their skill in making long voyages and their confidence in turning away from the African shore to find an offshore route to the Indian Ocean. The heir of the accumulated geographic knowledge of Portuguese mariners, Vasco da Gama, followed prevailing winds from Lisbon to the Cape Verde Islands off the westernmost point of Africa, and then continued directly to the Cape of Good Hope. He did not pause at Elmina (São Jorge da Mina) in present-day Ghana, the first trading post on the Gold Coast, which soon became an important site for the Atlantic slave trade. This route became the standard one from the Atlantic to the Indian Ocean for centuries. Skipping the stop at Guinea allowed navigators to keep away from the coast and signalled the new goal of reaching India and competing for the spice trade.

Vasco da Gama's success at entering the Indian Ocean in 1497 and crossing it depended in large part on the expertise of a knowledge-able Muslim pilot he met in East Africa before attempting the final leg of his voyage. Success by European explorers frequently rested on the local skills and knowledge of people whose cultures had sustained ties to the ocean, such as the Indian Ocean navigators who knew how to manage the trade winds to traverse the seas between Africa and China. Europeans also sought knowledge of these waters from Asian maps, such as a Javanese chart shown to a Portuguese captain in 1512, as well as from expert sailors.

The fleet da Gama commanded included one caravel and three ships. Until modern times, 'ship' was not a generic term but referred to the specific characteristics of a vessel that carried three masts, all with square rigged sails (square to the centre-line of the boat, but really more rectangular in shape). Caravels were smaller than ships, and lateen-rigged – that is, with triangular sails oriented along the long axis of the ship. This fore-and-aft rig served well for coasting voyages involv-ing sailing upwind, but each triangular sail was very large and required many men to handle. Square rigs proved superior for sailing with the wind, a critical quality for vessels following trade winds across oceans. Square rigs also had several sails on each mast, each one small enough to be handled by fewer sailors than large triangular sails of lateen-rigged vessels. Importantly, ships had greater capacity than caravels for carrying provisions, water, guns and trade goods in addition to large crews.

The Portuguese discovery of a sea route to China and its desirable trade goods supports Parry's argument about discovery of the sea, but what of Christopher Columbus's accidental and wholly unexpected discovery of a New World instead of the Far Eastern islands he sought? Columbus was a skilled and painstaking navigator from the Italian city-state of Genoa who believed, based on information he carefully compiled, that he could reach the east by sailing west. Columbus's sources included rediscovered ancient Greek geographies, including Ptolemy's depiction of the Earth as a sphere, as well as the accumulated experience of Atlantic voyagers, from whom Columbus learned of the Sargasso Sea. Discovery of the Canaries, Azores and Madeira seemed to confirm lore about lands and islands lying to the west. Medieval stories of the island of St Brendan or the islands of the Seven Cities provided reassurance for mariners headed into the otherwise open and frightening Atlantic. For crossing the high sea, Columbus employed a navigational technique known as dead reckoning (from 'deduced reckoning') that involved following a steady compass heading and keeping track of the estimated distance sailed. Columbus believed the distance west to China was smaller than it is, based on his misjudgement of the circumference of the Earth, but this error only serves to emphasize his goal of finding a sea route for trade with the East.

History has traditionally emphasized Columbus's achievement as bringing together the long-isolated continents of North and South America with the Old World. The Columbian Exchange of people, animals, plants and pathogens indeed indelibly altered the planet. Columbus himself, though, remained convinced that he had reached a land known, not unknown, to Europeans. The explorers who followed him did not rush to explore the land Columbus found. Remaining intent on the Orient, his successors first concentrated on looking for a way around it. Explorers sought a passage among the Caribbean islands, through the Chesapeake, in the icy waters of the northwest of North America and, as Ferdinand Magellan finally did with success, by sailing south to skirt the inconvenient landmasses blocking the preferred westerly tropical or temperate routes west. Columbus's crucial contribution to the discovery of the sea was not bumping into the New World but rather rendering the vast and terrifying Atlantic Ocean finite and bounded.

The activities of the Spanish and Portuguese in the Atlantic offer insight into Western understanding of the ocean at the time. The two maritime powers became locked in a geopolitical struggle over control of long-distance trade and turned to the Catholic Church for adjudication. Papal bulls and treaties etched an imaginary line from the North to the South Pole west of the Cape Verde Islands, giving Portugal exclusive rights to explore the African coast and to seek an eastern route to India, reserving the western Atlantic and a possible western route to the Far East for Spain. It would be easy to interpret the 1494 Treaty of Tordesillas as a division of the world, or of the ocean itself, as some commentators have done. However, it is more accurate to think of the agreement as an allocation of two directions of exploration rather than a rendering of the ocean as sovereign territory. The ostensible motive for Portuguese and Spanish exploration, endorsed by the Catholic Church, lay in spreading the Gospel, an activity that could not apply to the uninhabited ocean. The sea served only as a transport surface to enable exploration, missionary activities and trade. It was possible to exercise social control in ocean space through the exclusion of other powers from exploring activity in designated areas, but the ocean itself was not claimable territory, as land was.

The culmination of the sixteenth-century discovery of the sea was Magellan's circumnavigation (1519–22), demonstrating a navigable southern sea route circling the world. Magellan (originally named Fernão de Magalhães), who sailed on behalf of Spain although he hailed from Portugal, was one of the many expert mariners who, as Columbus had, pursued personal interest by serving a foreign monarch. The expedition, though a geographic success, limped home. Three of the five ships were wrecked and almost two hundred people including Magellan himself lost their lives. Yet emerging from the human disaster came optimism for future trade and important new knowledge about the globe. The world proved larger than many expected. The most astonishing discovery was the Pacific, a third great ocean lying between the Americas and Asia. Although geographic questions remained, for example about the possibility of a southern continent or the persistent dream of a Northwest Passage, sea routes between all known places had been found.

AN ACCOUNT OF MAGELLAN's circumnavigation appeared in print just one year after the expedition's return, joining a growing body of published material about geography and seafaring. Johannes Gutenberg printed his first Bible in 1455, and the first published sailing directions appeared 35 years later. Print media encouraged the divergence of navigational information from material discussing the commercial prospects of trade at various ports. Printing promoted the widespread distribution of geographic and hydrographic information, including maps, to readers throughout Europe at a time when literacy was on the rise and the spreading use of vernacular languages made such works available to non-scholars.

Discovery of the sea leaned heavily on geographical and maritime information available from learned sources, including Arabic texts and rediscovered Greek texts. The Renaissance rediscovery of classical Greek learning included geographic writings of Herodotus, Aristotle and Ptolemy, each of whom rejected the traditional view of an oceanic river flowing around a circular world. Ptolemy's *Geography*, restored to Europe from Byzantium early in the fifteenth century, countered the Earth island idea with a view that oceans were entirely separated by landmasses, a model challenged by Magellan's achievement of sailing between oceans.

Europe's explorers actively sought and exploited both academic knowledge and geographic experience in their systematic search for new trade routes. Use of the sea ultimately rested on reliable knowledge of the ocean. Fresh appreciation for empirical evidence fuelled recognition of the value of experience, and the process of exploration included mechanisms for accumulating and disseminating new geographic knowledge to form the basis for future navigation.

At the outset of the discovery of the seas, portolan charts recorded actual experiences at sea. These navigational aids provided mariners with compass direction and estimated the distance between coastal landmarks or harbours. Utterly novel for their time, portolans were the first charts to attempt to depict scale. Portolans created by fourteenth- and fifteenth-century explorers document Portuguese and Spanish discovery of Atlantic islands and the African coast, and helped subsequent mariners retrace their steps. Accuracy of portolans was best over shorter distances, and they became less useful when navigators steered offshore.

In contrast to creators of portolans, armchair cartographers compiled world maps of little use for actual navigation but which reflected shifting

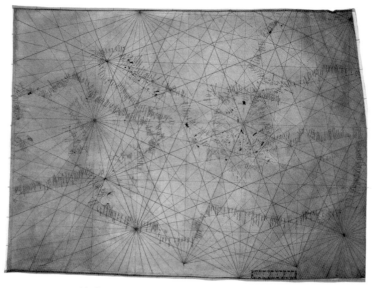

Mediterranean portolan chart from the 14th century.

knowledge of oceans. While manuscript maps had been produced alongside written manuscripts since antiquity, the earliest known printed map was included in an encyclopedia of 1470. It represents the world schematically within a circle, in which the three continents of Asia, Europe and Africa are surrounded by an ocean river and separated from each other by horizontal and vertical rivers that form a T shape; hence the name 'T-O' to describe this kind of map. Other early maps were based on Ptolemy's work, on biblical stories or other allegories, or occasionally on portolans.

Cartographers created maps for patrons or in hopes of selling printed copies, generally seeking recent information about the ocean from a wide variety of sources including published maps and books, manuscripts, reports from mariners and other voyagers and, perhaps in a few cases, personal experience. Medieval bestiaries, or illustrated compendia of animals, provided models for sea monsters that appeared on maps, but Renaissance cartographers also borrowed from classical images, such as stylized dolphins from ancient Rome. Legends such as the tale from St Brendan of running aground on an island that turned out to be the back of a whale inspired pictures on maps of mariners making fires on the backs of enormous animals mistaken for oceanic isles. Medieval maps that included sea monsters or other astonishing or exotic marine creatures often portrayed these as oceanic versions of terrestrial animals, such as sea dogs, sea pigs or sea lions, following an ancient theory that the ocean harboured animals equivalent to those on land.

Although the majority of medieval maps and nautical charts of the Age of Discovery did not include sea monsters, the ones that do reveal both a rise of general interest in marvels and wonders and a specific concern for maritime activities that took place at sea, including in far-distant oceans. The more exotic creatures are often positioned on maps at the edge of the Earth, conveying a sense of mystery and danger and perhaps discouraging voyages in those areas. Images of octopuses or other monsters attacking ships would seem to be warning of dangers to navigation. A striking depiction of King Manuel of Portugal riding a sea monster near Africa's southern tip, from Martin Waldseemüller's 1516 *Carta Marina*, displayed Portuguese mastery and political control over the sea.

The most famous representations of sea monsters in the sixteenth century frolic on the *Carta Marina* created by the Swedish priest Olaus Magnus and published in 1539. The map covers Scandinavia south to northern Germany, east to Finland and including the Atlantic to the south and slightly west of Iceland. Land is represented with much detail, but so too is the ocean. There are real as well as non-existent islands, and ships carrying cargo or fishing. The sea is full of dynamic marine creatures, some recognizable as whales, sea lions, lobsters or other species of economic value but others more fantastic or menacing. Monsters attacking ships from Protestant countries may reflect the exiled Magnus's displeasure at reformation after Sweden broke from the Catholic Church.

In addition to his map, Magnus also wrote detailed descriptions of his sea monsters in his book, whose translated title is *History of Northern Peoples* (1555). While some of his marine creatures derived from bestiaries or legend, such as his account of a 60-m (200-ft) sea serpent living

Martin Waldseemüller's world map *Carta Marina*, with the King of Portugal riding on a sea creature near the southern tip of Africa, 1516.

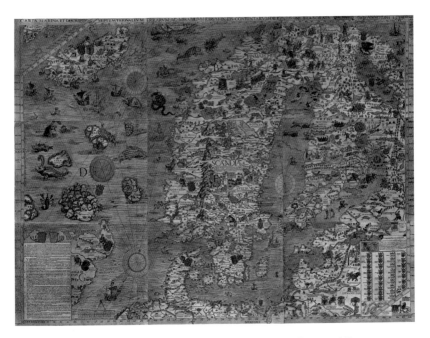

Carta Marina by Olaus Magnus which, between Iceland, the Faroes and Norway, shows whorls that modern oceanographers believe refer to the front where the warm Gulf Stream waters mix with cold currents moving around the coast of Iceland.

near Bergen, Norway, others of his depictions were original, and some may reflect reports of unfamiliar animals. Mariners returned home with stories of strange creatures they saw at sea, including giant whales and sharks, as well as narwhals with single, helical tusks up to 3 m (10 ft) in length. Until explorers discovered the truth, narwhal teeth were sold in Europe and the Far East as unicorn horns. The European court knew the ocean to be the source of strange and precious objects such as coral, pearls and mollusc shells, which began to appear in seventeenth-century cabinets of curiosities alongside dried sea horses, turtle shells and shark's teeth. The Swiss naturalist Conrad Gessner assembled existing knowledge about marine life in his 1558 *Fisch-Buch* (Fish Book), which listed about eight hundred creatures. He drew upon mariners' tales, requested drawings and specimens from sources throughout Europe, and travelled to the Venice fish market to study for several months.

Illustration by Swiss naturalist Conrad Gessner of a whale attacking a boat, 1560.

Embracing the variety of living forms as proof of divine creation, Gessner's work included both real and mythological creatures. His *Historia animalium*, published over the course of the 1550s, featured a unicorn among the terrestrial animals, while his sea animals included mermaids, a walrus with feet, and a number of monsters taken from Magnus's work. Numerous other authors and cartographers similarly copied or borrowed Magnus's influential images.

The detail captured and information conveyed in the ocean spaces of many maps testified to the extent, liveliness and importance of maritime activity and its oceanic setting. Representations of marine creatures and other cartographic choices revealed to the viewer of a map what was normally hidden, rendering the ocean as a thoroughly three-dimensional space full of life and activity. Magnus depicted the sea's surface as highly textured by filling all ocean space with horizontally oriented dotted lines, occasionally interrupted by wake from a moving vessel or a surfacing animal. Textured surfaces, identifiable marine creatures or exotic and forbidding monsters, and ships busily prosecuting maritime work fill map space devoted to the ocean on many cartographic works produced through the sixteenth century. The amount of such decorative detail on a map might depend on the amount of money a patron was willing to spend.

Medieval maps published in the first fifty years after the printing press often included textured oceans, represented sometimes with parallel lines and other times with broken, wavy lines. Some allegorical maps likewise featured oceans filled with lines denoting a watery surface, calling attention to what might lie beneath, while other maps that divided ocean space by latitude and longitude left oceans blank but used sea creatures and ships as decorative elements symbolizing maritime activity.

One detail on the Magnus map would seem to speak to navigators. In the area east of Iceland and extending to north of the Faroe Islands, there is a band of carefully drawn whorls, which modern oceanographers argue may represent the Iceland–Faroes front. Here, cold Arctic water currents coming around the coast of Iceland mix with warmer water from the Gulf Stream flowing north to form strong eddies that could drive a ship off course. Trade conducted between Iceland and the European continent by mariners of the Hanseatic League brought dried cod and other products from Iceland to exchange for grain, beer, wood and textiles. Navigators certainly noticed temperature differences and also variations in surface currents, and this information was likely conveyed to Magnus during visits to northern German cities. One other map of Scandinavia from Brussels, made in the late fifteenth century, may portray this same oceanic feature also with a series of whorls. In neither map are there such curls shown anywhere else in the ocean, suggesting they were meant to represent a specific feature, one no doubt long known by Hansa navigators and Vikings before them.

Maps became more useful to navigators when cartographers created projections that could represent the spherical Earth on the flat, printed page. For open-ocean journeys, long-distance voyaging navigators turned to the Mercator projection (1569), which showed the latitude and longitude grid and corrected for the Earth's curvature such that lines of a compass heading could be represented on the chart as straight lines or, as sailors called them, rhumb lines (loxodromes). This made it possible to keep track on the chart of a route sailed by dead reckoning. With the addition of celestial navigation, mariners could use observations of the height of stars or the Sun taken with quadrants to find their latitude and correct mistakes that arose from magnetic variation or dead reckoning navigation. Mariners would not gain the ability to find their longitude until the late eighteenth century.

Knowledge recorded on Mercator projection charts prioritized actual observations, soon to be prized by scholars associated with changes wrought by the Scientific Revolution, generally understood to begin with Nicolaus Copernicus and the publication just before his death in 1543 of his model of the universe with the Sun rather than the Earth at the centre. Cartographer Gerard Mercator himself sought the newest and most trustworthy sources of information for his work, shifting from copying Magnus's sea monsters in the terrestrial globe he created in 1541 to using recent natural history publications for his subsequent famous 1569 world map. He moved his most exotic creatures from the regions around Africa, Asia and the Indian Ocean, where most cartographers before then had placed sirens and other dangerous or mysterious monsters, to waters off South America and the Pacific Ocean, signalling the rise of geographic interest in these relatively less-travelled parts of the globe.

A friend and rival of Mercator, Abraham Ortelius, created a world map in 1570 that included elements by then common in visual representations of the ocean as well as some newer features. He textured the sea's surface with stippling over all of the ocean space on his map, but he also divided that space with lines of latitude and longitude. There is one ship and two creatures that might be whales, possibly decorative elements but certainly also acknowledgement of maritime activity and a reference to the importance of whales as resources. Other maps from later in the sixteenth century feature images of lifelike whales and of people whaling, including one of the North Atlantic and what is now

Canada with an inset drawing of men, possibly Basques, harpooning one whale and flensing another. This scene was copied from the first published print of Basque whale hunting at sea that had been produced a decade earlier. This map also included whale-like monsters resembling sea monsters from older sources, but the trope of sea monsters on maps and charts was fading.

In an increasingly well-known world, monsters on maps made after the sixteenth century appeared to convey fancy or whimsy rather than danger. Cartographers began to depict economically valuable marine animals realistically based on up-to-date natural history fostered by the growing experience of mariners. Actual observations of nature replaced preoccupation with marvels and wonders. Whales on charts transformed from threatening giants into swimming commodities and then disappeared entirely. The occasional decorative ship, compass rose or nautical instrument on maps gestured to human maritime activity, but banished the ocean itself. The gridded ocean space of the seventeenth century and beyond was most often left blank, without stippling or lines to denote

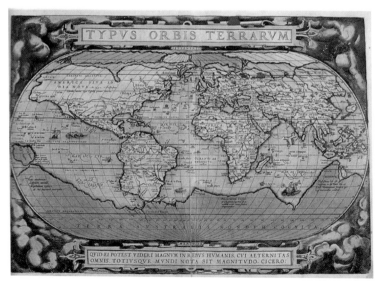

Abraham Ortelius's world map, *Typus Orbis Terrarum*, 1570, by which time sea monsters had been relegated to the yet-unexplored places on Earth.

the presence of the sea itself or to conjure the world hidden beneath its surface. Blank ocean space on maps appears to reflect a revision of the sea from a dangerous and mysterious place into a knowable part of the natural world whose control enabled the exploitation of marine resources and the expansion of European power.

PEOPLE CAUGHT UP in the European discovery of the sea recognized the joint contributions of the printing press and the magnetic compass to the weaving of global power. To these mechanical discoveries, the statesman, jurist and essayist Francis Bacon added gunpowder, crediting the trio of inventions with altering the world more than any empire. Of greater importance than any single one of these technologies, Bacon understood science – the creation of new knowledge about the natural world from empirical investigation – as the source of power. The frontispiece of the 1620 philosophical work in which he celebrated the compass, printing press and gunpowder linked Bacon's creation, the scientific method, to exploration. It depicted a ship sailing out from the Mediterranean into the Atlantic Ocean through the mythical Pillars of Hercules flanking the Strait of Gibraltar. To Bacon, who is credited with the aphorism 'knowledge is power', science and exploration created knowledge and thus conveyed power.

Bacon at the time was a citizen of the burgeoning maritime state of England during a period when northern European states challenged Iberian power. Portugal and Spain tried mightily in the sixteenth century to exploit their respective eastern and western routes to the Far East, but shipping did not completely dominate overland routes and trade conducted along them by well-established Arab and Venetian powers until the following centuries. Nor did the emerging maritime empires of the Dutch, English and French meekly abide by Iberian efforts to enforce their exclusion from the ocean routes.

Early attempts by northern European nations to locate a northern passage to Asia found instead fish (and, also, via Hudson's Bay, plentiful

The frontispiece of Sir Francis Bacon's *Instauratio Magna* of 1620 features a ship sailing through columns that denote the Strait of Gibraltar, departing the world of classical learning for the open ocean, symbolizing limitless natural knowledge.

furs in the northern terrestrial wilderness of North America). The rich North Atlantic fisheries fuelled competition for ocean trading routes by providing a valuable commodity, not as lucrative as sugar or other sources of wealth from the New World but a critical export for a region that, unlike the Caribbean, attracted many northern Europeans to settle. Long before permanent colonies, northwest Atlantic fisheries attracted first the Vikings and the Basques and subsequently the English and other northern Europeans. English fishers began to seek cod beyond their own waters in seas around Iceland in response to the Hanseatic

League's monopoly over fisheries and the trade for this important source of protein for observant Catholics.

To European eyes, not recognizing the severe depletions in their own waters from fishing, the abundance of marine life in unfished areas was overwhelming. John Cabot, like Columbus a Genoese mariner, originally named Giovanni Caboto, led the first English expedition to North America in 1497. He wrote of being able to catch cod simply by letting down a basket, while stories circulated of shoals of cod so thick they slowed the progress of vessels. By 1575 more than 300 vessels from France, Portugal and England fished the Grand Banks and by the end of the century, that number rose to about 650, harvesting thousands of tonnes of fish. Trade in salt cod helped knit together the maritime economies of northern and southern Europe and stimulate transatlantic trade.

Hardly content to use the ocean only for its cod, the English, French and Dutch asserted networks of trade beyond Europe and the North Atlantic, and began to develop navies. Maritime conflict shifted from efforts to protect or attack shipping for commercial, or even personal, advantage to political conflicts between European states asserting their mercantile interests in trade routes, entrepôts or colonies. After the launch of the warship *Mary Rose* in 1510, England embarked on the creation of a state navy, and other European powers followed suit. Elizabeth I ascended the throne of England in 1558 and almost two decades later dispatched Francis Drake on a series of improvised voyages that became the world's second circumnavigation. Considered a pirate by the Spanish, whose ships and outposts he attacked, and a privateer by his queen, Drake returned with captured treasures, was rewarded with a knighthood and served as second in command in the English fleet that fought the Spanish Armada in 1588. Its defeat, while not ending conflicts with Spain, did strengthen England and demonstrate to all of Europe that Spain was not invincible.

Trade and warfare at sea remained inextricably connected as northern European powers challenged the Iberian monopoly on ocean routes to eastern markets. Elizabeth's adviser Sir Walter Raleigh articulated the direct link between control of the sea and world power in an essay written around 1615, but published after his death, declaring,

> He who commands the sea, commands the trade routes of
>   the world.
> He who commands the trade routes, commands the trade.
> He who commands the trade, commands the riches of the
>   world,
> and hence the world itself.[1]

In 1600 English merchants had formed the East India Company, a joint stock company. Two years later the Dutch government chartered the Dutch East India Company (Vereenige Oostindische Compagnie, or 'VOC'). When the Dutch company captured a Portuguese ship in Asian waters, it engaged the jurist Hugo Grotius to defend its right to seize the cargo and the ship. His 1608 pamphlet *Mare liberum* argued that historically the seas had been free for all to use.

Mare liberum set off what international jurists have dubbed the Battle of the Books, a debate between Grotius and scholars from Portugal and England who responded to his contentions. Grotius rejected the possibility that Portugal could possess the ocean for several reasons. Scoffing at the imaginary line that demarcated Iberian claims to the ocean, Grotius noted that people cannot inhabit the ocean or divide it into territories, as they can land, and that the inability to bound the ocean renders it un-claimable. He dismissed Portuguese claims to discovery of the sea by recognizing prior use of all seas by the people who inhabit adjacent coasts, citing Moors, Ethiopians, Arabians, Persians, peoples of India and others. Importantly, he contended that the two primary uses of the ocean, navigation and fishing, are inexhaustible. Natural law, he concluded, dictates that use of the ocean's waters be open to all. Since the ocean exists as an arena where states interact, if a state should gain the ability to possess the sea fully, that state would retain a responsibility to allow access to other users.

Grotius's arguments and the counterarguments mounted against them reflect the maritime policies of expansionist states and the conceptions of the ocean that followed from these. A Portuguese response to *Mare liberum* agreed to the principle of the ocean as un-possessible space but asserted the possibility of exclusive usufruct rights over particular trade routes, thus defending the Treaty of Tordesillas. Competition between the Dutch and English, not only over long-distance trade but over fishing in their respective coastal waters, informed the work of the English jurist John Selden, who defended claims for English sovereignty over as much of the sea as Britain could defend. Selden's argument can be understood as a variation of Grotius's in that his claim for a state's possession of nearby ocean space was accompanied by recognition of some rights to common usage. And he essentially acknowledged the deep sea as free space by not addressing the region beyond the national waters he claimed for England.

Selden wrote the treatise *Mare clausum* in 1619 but it was not published until 1635, during a push by Charles I to rebuild the navy in hopes of policing English coastal waters. During the seventeenth century, trade across oceans resulted in the proliferation of far-flung European settlements and commercial outposts in the Americas and Asia. While the concept of freedom of the seas appealed to northern European powers asserting the right to trade in Asia, commercially motivated conflicts among European powers over access to those markets contributed to five major wars during the second half of that century. Three Anglo-Dutch wars between 1652 and 1674 forced the Netherlands to accept English claims to sovereignty in its own coastal waters and weakened the Dutch Empire, leaving France as Britain's strongest competitor for colonial supremacy thereafter.

In practice, European powers' embrace of the concept of freedom of the seas was inconsistent and based on expediency. At least initially, northern European powers imitated their Iberian predecessors in attempting to exert monopolies in the maritime realm. The Dutch had championed the idea of free sea, but pursued monopoly in the Indian

Ocean region, while the English had argued for open seas there yet tried to exercise control over adjacent seas near home. European encroachments in Asia encountered local resistance, as when opposition to Dutch efforts to monopolize trade in Madagascar in the seventeenth century prompted a defiant declaration that echoed the arguments of Grotius: 'God has made the Earth and the sea, has divided the Earth among mankind and given the sea in common. It is a thing unheard of that anyone should be forbidden to sail the seas.'[2] In fact, Grotius's articulation of freedom of the seas was informed by his awareness of the long and undisputed tradition of freedom of navigation and trade embraced by Asian states and Arab navigators in the Indian Ocean region.

Outside legal writings, the ocean – including the possibility of control over it – featured prominently as the setting for plays and popular writings. William Shakespeare's *The Tempest* (1611), for instance, displayed the ocean resisting civil authority and therefore possession. Maritime activities were the subject of accounts of notable voyages and dramatic shipwreck narratives but also of pilot guides, practical manuals of seamanship, atlases and other works intended for a professional audience of mariners, politicians and government officials, merchants and entrepreneurs, and scientists and engineers. Though this body of maritime literature served a practical function, it also thrilled a new audience of readers: armchair sailors.

Global exploration greatly expanded knowledge of the ocean, but Westerners who were not mariners perceived the sea simply as a wilderness, a place outside human control. Jonathan Raban, editor of *The Oxford Book of the Sea* (1992), noted with surprise when writing about the voyaging literature of the seventeenth century, 'there is so little sea in it.'[3] The literature of the English Renaissance is filled with maritime voyages, instruments, approaches to land, encounters with other vessels, and events and people on board ships. On the rare occasions when the sea itself does appear, it is almost always because of a horrific storm. William Bradford's comments about the 1620 *Mayflower* crossing mentioned 'perils and miseries' as the vessel sailed 'over the vast and furious ocean'.[4] A decade before, a colonist headed for Jamestown chronicled the experience of his vessel, the *Sea Venture*, in a hurricane. He wrote, 'Greater violence we could not apprehend in our imaginations. Winds and seas were as mad as fury could make them.'[5] The ocean shared many of the same characteristics as terrestrial wilderness in European perception. Bradford called the area where his party landed 'a hideous and desolate wilderness' yet revealed his unqualified preference for wild spaces on shore in his fervent thanks to God for letting them '[set] their feet on the firm and stable Earth, their proper element'.[6]

Europeans migrating to North America, by then understood as the New World, were part of the large-scale migration, both voluntary and involuntary, which began in the sixteenth century and accelerated as the maritime systems that had been perfected to carry cargo were adapted to transport people in the eighteenth century. Altogether, about 11.4 million people travelled to the Americas in sailing vessels between 1500 and 1820; the vast majority – more than three-quarters – were enslaved Africans. Among these were people from West Africa who transported swimming and diving skills to the New World. Western travellers between the sixteenth and nineteenth centuries frequently noted the swimming abilities of West Africans, who used variants of the freestyle stroke (as did indigenous people in the Americas and Asia) at a time when most Europeans could not swim. Enslaved West Africans who were carried to the Americas, when they lived near waterways, swam for recreation and taught their children to swim. Some slaveholders took advantage of slaves' diving skills by putting them to work as lifeguards or employing them in salvage diving, pearl harvesting or clearing riverbeds for fisheries or navigation.

In step with the expansion of human transportation across oceans and increases in shipping prompted by the rise of industrial capitalism, the numbers of people who worked aboard ships increased dramatically. In addition to the migrants, there were merchants; fishers; whalers; sealers; the crews of cargo, slaving and passenger vessels; and also the officers and crews of naval ships. Those who worked at sea experienced the ocean as a space of human activity, but increasingly so did writers, cartographers and readers.

The naming of parts of the ocean suggests that before the nineteenth century the ocean was widely viewed as connected to people and their endeavours. Magellan gave the Pacific its first unitary name, 'peaceful sea', without knowing its enormity or violence. But often during the eighteenth century, Westerners named subsections of the Pacific and other oceans after their adjacent lands or after linkages between basins, revealing an understanding of the sea as a space of human activity. Examples include the 'Barbarian Sea' off East Africa or the 'Chinese Ocean' of the western Pacific, as well as the seas of Peru, Brazil or Chile, whose names evoked the routes and exchanges between these places. The waters between Africa's horn and the islands of the eastern Indian Ocean that supported slaving and plantations were called the 'Ethiopian Sea'.

The maritime literature of the eighteenth century likewise stressed human activity at sea and on the ocean's margins. Daniel Defoe set his 1719 story about the shipwrecked Robinson Crusoe in the Pacific. *Robinson Crusoe* is a fictionalized version of a true story about Alexander Selkirk, a mariner rescued from the island of Juan Fernández off the coast of Chile. Defoe's contemporary and inspiration, the buccaneer, captain and explorer William Dampier, rescued Selkirk during his third circumnavigation, a feat Dampier was the first person to accomplish. His 1697 adventure narrative, *New Voyage Round the World*, published after his first circumnavigation, provided invaluable information on winds, currents and coasts, as well as on the natural history and people of unknown places. The book garnered respect from men of science and sold almost as well as Defoe's popular tale. Defoe's work initiated a few decades in which maritime stories celebrating the craft of seamanship were extremely popular, until historical accounts of voyages eclipsed fictional ones in the second half of the century, when Pacific exploration produced many compelling narratives. Expedition stories shared readership with sensationalized pirate biographies and the ever-popular shipwreck tales, as well as accounts of the harsh realities of life as a common sailor and a few narratives by black seamen.

Maritime writings accentuated human endeavours and the skill and knowledge required to operate at sea. Unrecognized went the results of that work in terms of impacts on marine populations and ecosystems. The acquisition of the island of Mauritius by the Dutch East India Company in 1598, for example, laid the foundation for the extinction of the dodo, a flightless bird that was hunted for food by sailors whose vessels came to the island for provisions. Human hunting and also predation by domesticated animals introduced by the small community the company established eliminated the species within a century, although its extinction was unrecognized until the nineteenth century. The dodo's status as the paradigmatic example of extinction derives from its appearance as a character in Lewis Carroll's 1865 children's story *Alice's Adventures in Wonderland*.

Maritime exploration was also responsible for the extinction, and near extinction, of marine animals as well. In 1741, large, herbivorous sea cows were discovered on the Commander Islands by the expedition led

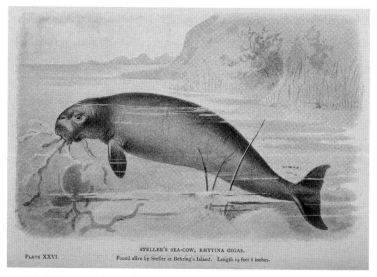

STELLER'S SEA-COW, RHYTINA GIGAS.

PLATE XXVI.   Found alive by Steller at Behring's Island. Length 19 feet 6 inches.

Steller's sea cow, featured in an 1896 popular book about extinct large animals.

by Vitus Bering and described by the naturalist Georg Wilhelm Steller, for whom the species was named. When the expedition discovered the remnant population, the once-abundant Steller's sea cows were already extinct in most of their original range across the North Pacific from Japan to Mexico, likely through aboriginal hunting. Seal hunters, fur traders and sailors followed Bering into the area and within 27 years hunted the mammals to extinction for food and fat and to use their skin to make boats.

Another marine mammal also first described by Steller found its species threatened by hunters that pursued it across the same northern Pacific range where the Steller's sea cow once flourished. Russian hunters, usually employing native Aleuts, pursued sea otters for their luxurious, incredibly valuable fur. Indigenous hunters had historically caused periodic local reductions, but commercial hunting extirpated populations first in the north and then farther south. In addition to almost eradicating the species, market hunting also caused cascading effects in the kelp forest ecosystem, when the otter's preferred prey, sea urchins, multiplied and grazed on kelp, causing the collapse of kelp forests, which in turn affected the numerous species that relied on them.

Exploitation of marine resources shifted from subsistence use to commercial activity in different parts of the world at different times. The apparent limitlessness of marine resources everywhere outside heavily exploited European waters, bolstered by the logic of free seas and the demands of mercantile capitalism, encouraged a pattern of serial depletion of local resources followed by movement to new places to hunt and fish. Pacific waters first felt the brunt of global markets in the eighteenth and nineteenth centuries as fur hunting, whaling, sealing and fishing began removing animals from that ocean, transforming them into commodities and circulating these in the global marketplace.

Between the sixteenth and eighteenth centuries, maritime trade supported states and expanding empires. Whereas Spain and Portugal had tried to use military might and the authority of the Catholic Church to exclude other states from the oceanic trade routes they considered their own, the way that the Dutch and English maritime powers used the sea suggested advantages for them and other European powers to the doctrine of freedom of the seas. The proliferation of overseas settlements and commercial enclaves in the Americas and Asia eventually prompted recognition that the sea was not a private fiefdom but a commons held by all. With Britain's global dominance, emerging over the eighteenth century, the doctrine of freedom of the seas was exerted or accepted around the world, including by former European rivals who recognized the benefit to them of the British defence of free seas. A 3-nautical-mile territorial sea, the remnant of *Mare clausum*, was recognized by most, but not all, nations. This was the distance a cannon shot could be fired and thus a shoreline defended. Free seas went hand-in-hand with the arrogant assumption that the ocean and its resources should be exploited by those with the knowledge and power to do so.

FREE TRADE AND FREEDOM of the seas prompted governments, particularly of northern European maritime powers, to replace private interests in managing transoceanic activities. Among the most important efforts by states to control the ocean, according to historian Joyce Chaplin, were parallel and related campaigns to solve the stubborn problem of longitude and to prevent scurvy on voyages lengthened by the inability to determine a ship's longitude at sea. Most deaths on Magellan's circumnavigation were caused by scurvy, now understood to be triggered by a deficiency of vitamin C. At the time, mariners recognized bleeding from gums and the loss of teeth, proceeding to wounds that wouldn't heal, as signs of impending death from scurvy, although finding land and fresh food sometimes reversed the progression of the disease. During the three hundred years that Europeans explored the Pacific but were unable to find inhabited islands or to settle and control the Pacific in ways they did elsewhere in the world, experts and sailors could not agree upon what measures prevented or cured scurvy. European mariners raided outposts of other European powers, as Drake did, or searched for native settlements to find what locals used to stay healthy.

The fortunate fictional travellers adrift in the Pacific and suffering from scurvy in Francis Bacon's story *New Atlantis* (1627) accidentally landed upon the island of Bensalem. Bacon placed in the most unknown part of the globe his utopian community that was organized to advance natural science and base governance on knowledge. The lengthy process of trying to conquer scurvy through experiment and theory reflected the European stance towards the Pacific in general as a place where science might form the basis for successful imperial control. This was in part a pragmatic response to the challenges presented by the Pacific as well as the sheer distance from Europe, which made reach by any significant naval force virtually impossible until the nineteenth century. The British navy did not systematically implement the use of citrus juice to fight scurvy until the end of the eighteenth century, nor solve the problem of longitude much before then, but the Age of Enlightenment opened with an effort by the British government to address the longitude conundrum.

The story of longitude parallels the consolidation of the British Empire. Shortly after the 1707 political union between England and Scotland that created Great Britain, the government offered an enormous cash prize, signalling the importance of the challenge, for an accurate method of determining longitude. Success rested on finding a means to know both the local time, easily measurable with navigational instruments, and the time at a location whose exact coordinates were known, such as Greenwich, England, or another of the contemporary competitors for location of a prime meridian, such as Lisbon, Madrid, Paris, Brussels, Antwerp or Rio de Janeiro. Most of the world's main sea routes had been found before navigators could determine longitude accurately. European knowledge of how to travel around the Pacific depended instead on information from indigenous peoples, often extracted by kidnapping individuals. The eventual triumph of

John Harrison's chronometers over the lunar distance method emerged after decades of work on, and improvements to, both prospective solutions, although ordinary navigators could not afford these timepieces until the nineteenth century.

Captain James Cook took a copy of Harrison's fourth chronometer on his second Pacific expedition in 1772 for testing. He also carried Dampier's books, relying upon his maps of trade winds, seasonal monsoons and equatorial currents and arriving at Australia almost eighty years after Dampier visited. Cook's voyages set a precedent for a tradition of scientific exploration subsequently led in Britain by Joseph Banks, a botanist who sailed on Cook's first expedition. Banks returned home to scientific fame, earning election as President of the Royal Society, a position he held for 41 years. From its helm, and through his role as informal adviser to King George III on the Royal Botanic Gardens at Kew, Banks orchestrated far-flung scientific collecting and expeditions on land and sea. His counterparts in the Netherlands, France, Russia and other nations, including the youthful U.S., likewise dispatched scientific exploring parties to the Earth's far corners.

By the end of the eighteenth century, conceptions of land and sea had shifted. New worlds had once been imagined as islands; stories and expeditions about islands just over the horizon had encouraged Europeans to venture out of the Mediterranean and into the open Atlantic. Discovery of the sea reorganized geographic understanding of the globe, so that continents rather than islands became the primary unit of land territory. Interiors of most continents, especially Africa, went unexplored by Europeans long after most of the world's coasts and islands were known and claimed by them. Until the twentieth century, coasting dominated seafaring, despite the emergence of human facility at crossing the blue waters.

As the Earth became defined by its largest landmasses, oceans were emptied of human activity in the minds of cartographers, poets and landscape painters, just as they became more important than ever for the extension of global empires. The pens of poets such as George

Gordon, Lord Byron, erased the sea as a setting for mariners' craft and transformed it into a sublime and romantic place. He wrote,

> Roll on, thou deep and dark blue Ocean – roll!
> Ten thousand fleets sweep over thee in vain.[7]

Likewise, the sea in the paintings of J.M.W. Turner from the early years of the nineteenth century became a natural environment of elemental power. The newly evacuated ocean was filled by Samuel Taylor Coleridge with fantastic creatures and powers described by his *Rime of the Ancient Mariner* (1798). Poetry and art portrayed a sea available for filling by the imagination, but the elemental ocean was, in fact, the product of historical change.

The newly emptied oceans on maps attracted scientific scrutiny and were slowly filled with the results of global tidal studies, investigations of geomagnetism, meteorological observations and bathymetric surveys. While the practical needs of navigators motivated scientific interest in the sea, the more natural philosophers learned about the ocean, the more they simultaneously experienced it as a realm of wonder and myth, as historian Natascha Adamowsky argues.[8] Specimens of crinoids, or sea lilies, and ammonites from the West Indian Ocean marked the first discovery of marine species formerly known only as fossils, generating speculation that the sea might still contain primordial life of the sort being discovered in the rocks by geologists. Marine bioluminescence, long familiar to fishermen, sailors and coastal dwellers but never perceived as particularly mysterious, intrigued natural philosophers as they began to learn about the ocean.

Similarly, sea monsters refused to vanish. A sea serpent appeared in the New World, off Cape Ann in 1639, and a Norwegian bishop reported a snake-like kraken in his 1775 natural history of Norway. A particularly exhibitionist sea serpent cavorted off Gloucester in 1817 and again in 1819, on one occasion witnessed by over 300 people. Though energetically hunted, the monster was never caught and the

Painting of the sea serpent sighting by the captain and officers of HMS *Daedalus*
in August 1848, whose report of observing it for twenty minutes caused a sensation
on both sides of the Atlantic.

sighting never explained, although many others were, or have since been, explained as misinterpretations of actual marine creatures. William Hooker, director of Kew Gardens and father of the accomplished botanist, explorer and defender of Darwin's theory Joseph Hooker, considered sea serpents a real possibility, as did other serious men of science. Louis Agassiz, the famous Swiss ichthyologist who emigrated to the U.S. in the mid-nineteenth century, claimed he could 'no longer doubt the existence of some large marine reptile allied to Icthyosurus and Plesiosaurus'.[9] No specimens were caught that might have allowed naturalists to identify or dismiss sightings, some of which were reported by respectable and qualified observers. Another instance of reliable witnesses sighting a sea serpent resembling the Gloucester one happened in the South Atlantic in 1848, when the captain and officers of HMS *Daedalus* watched an 18-m (60-ft) dark-brown creature swim towards, and then nearby, the ship for twenty minutes. This sighting, too, remains a mystery.

Efforts to comprehend the sea resulted from the expansion of maritime trade and the increasing interest of governments in protecting that activity from dangers posed by the natural world. At the end of the eighteenth century, the British government founded the Hydrographic Office as a source of reliable charts formerly available only through its Dutch and French rivals. In the 1830s and 1840s, the u.s. likewise invested in institutions to promote maritime commerce. Government patronage of maritime affairs included some of the earliest state funding of science in order to increase knowledge of the ocean. Institutions such as the French Hydrographic Service, the American Naval Observatory and the Royal Dutch Meteorological Institute transformed from clearing houses for information into research and development operations.

The most powerful products of such institutions were charts, tide tables, sailing directions and other tools that enabled the movement of knowledge about the ocean from experienced mariners and men of science to publishers to the captains whose vessels acted as agents of empire. In England, for example, the long-time Hydrographic Office director, Francis Beaufort, vigorously promoted the use of the numerical wind scale that bears his name. It rendered mariners' experience with storms and calms into an easily communicated, quantified system so that observations could be compared globally and across time. Nations capitalized on new knowledge of the sea to gain a measure of control over it, not direct political power but the ability to use the ocean to weave a web of trade routes around the globe that tied together colonies and their raw materials and markets with industries in the home country.

Government support for science and the scientific study of the sea that nourished imperialism emerged conjointly. Indeed, the very term 'scientist' was coined by the physicist William Whewell in the context of his investigation of global tides. The way that Western imperial powers understood and used the sea increased the status of science in society.

Scientists, in turn, defined the ocean by measuring and recording its physical features, interpreting its laws, and creating charts and other representations of the ocean that could be used to extend imperial power.

BEGINNING IN THE FIFTEENTH CENTURY, European maritime powers found routes between all of the world's oceans. The enormously long story of connections between people around the globe and their adjacent seas appears in retrospect to serve as prologue to the dramatic changes wrought by the discovery of the sea. Long-standing uses of the ocean for trade, fishing and warfare intensified in scale and scope. Maps and books through the sixteenth and seventeenth centuries reveal a cultural perception of the sea as a site for human activity. Mariners, government officials and readers of maritime tales understood the ocean as a place full of wind, weather and currents that challenged navigation and equally as a place whose depths hid valuable resources and mysterious and unknown elements. Efforts by southern European powers to lay claim to sea routes were challenged by their northern neighbours whose defence of free trade ultimately supported free seas. The Scientific Revolution reinforced the emerging preference for experientially based knowledge over ancient learning or myth. Accumulation of knowledge about the sea supported the increasing scale and scope of its traditional uses. Governments embracing trade and colonization took measures to protect and promote these extensions of power, including the innovative step of patronizing science. Knowledge of the ocean won from work at sea was joined by knowledge produced by modern science. Together, traditional experience and new understandings of the ocean banished sea monsters to the edges of the known world, leaving blank ocean space on maps available for inscriptions conveying useful information to imperial agents. Science joined warfare as a tool for controlling the ocean through rendering the sea amenable to use by any power that could master its winds, currents and contours effectively. Freedom of the seas appealed most strongly to European powers with the infrastructures to know and use the open ocean and, increasingly, its depths. While knowledge undoubtedly enabled Western states to use the ocean to project power, imagination, in the form of ambition and desire, was equally implicated in the intersection of science, imperialism and oceans. Imagination continued in the nineteenth century to inspire new uses of the sea that would draw all of the ocean, including its greatest depths, into the sphere of human activity.

∾ FOUR ∾

Fathoming All the Ocean

Thou glorious mirror, where the Almighty's form
Glasses itself in tempests; in all time,
Calm, or convulsed – in breeze, or gale, or storm,
Icing the pole, or in the torrid clime
Dark-heaving; boundless, endless, and sublime –
The image of Eternity – the throne
Of the Invisible; even from out thy slime
The monsters of the deep are made; each zone
Obeys thee; thou goest forth, dread, fathomless, alone.
– George Gordon, Lord Byron, from *Childe Harold's Pilgrimage*,
Canto IV (1818)

T HE NINETEENTH-CENTURY discovery of the depths marked the start of the human relationship with all of the ocean, including its most remote reaches and most inaccessible parts. The ocean is a challenging place to know. Some of its characteristics – such as its opacity, tracklessness and vast scale – profoundly directed and constrained how people have amassed knowledge about it. Historically, sailors, navigators and fishers knew the ocean through their work, using trusted tools and hard-won knowledge passed down through generations. Traditional uses of the sea, especially fishing, trade, emigration and travel, intensified during the nineteenth century as industrialization transformed the blue water into a workplace on an entirely new scale.

New uses for the open sea and the depths that drew seafarers away from regular routes and familiar fishing grounds emerged. When whales became scarce near shore, their hunters embarked on longer voyages farther from land, pursuing deep-diving sperm whales. Whalers' stories sparked questions about the conditions at great depths and whether or not living things could exist there. Their experiences, tarrying in waters where no previous mariner had paused, inspired writers and engrossed readers. Submarine telegraphy, after successfully spanning short underwater routes, animated the ambition of engineers, entrepreneurs and politicians who dared to imagine transoceanic cables. Such new uses for the sea, far beyond the traditional ones of transportation and fishing, created unprecedented demands for knowledge about the ocean that governments met by supporting research. People before and through the nineteenth century learned about the ocean indirectly, employing technologies, skills and knowledge systems both from traditional forms of work at sea and from modern science, in order to make sense of the ocean's great extent and its profound depths.

More revolutionary than the increased scale of work at sea was the innovation of playing by and on the ocean. Beach holidays, yachting and visits to public aquaria provided salubrious and socially appealing access to the sea. The nineteenth-century discovery of the ocean extended into the private space of the home when families brought back collections of shells or seaweeds, tended aquaria or read maritime books. Alongside its significant and novel political and economic importance, the ocean gained tremendous cultural and even personal resonance.

NAVIGATORS BEFORE THE NINETEENTH century were more concerned to rule out shallowness than to measure depth with any precision. Careful measurement was reserved for waters near land. The 1823 *Encyclopædia Britannica* entry for 'Sea' stated that for 'want of proper instruments, beyond a certain depth, the sea has hitherto been found unfathomable.'[1] Standard navigational sounding equipment included only about 100 fathoms of line at most, while even explorers carried only 200 fathoms, and 'off soundings' referred to sailing in areas deep enough not to concern a navigator. One reason for this practice was rooted in the doubt about whether an object such as a sounding lead thrown overboard in deep water would ever reach bottom. Some mariners, and some natural philosophers as well, believed that water might be compressible, so that its density would increase with depth. If so, an object would float at the depth at which the water's density matched its own. Reports in the mid- to late nineteenth century record sailors' fear that telegraph cables or a shipmate's body committed to the deep might, as they termed it, 'find their level', and drift around indefinitely in mid-water column.

One distinctive feature of northern European navigation, relative to its Mediterranean origins, was its reliance on sounding. Pilots in the Mediterranean did not need to measure the depths often because its basin slopes steeply from the coasts, its shallow waters are clear, and fogs are rare. The Atlantic, by contrast, with its widely varying seabed slopes, frequent fogs and opaque water, prompted careful attention to the sea's third dimension. Mariners learned that the Atlantic's depth dropped off steeply past the 100-fathom mark, now recognized as the outer edge of the continental shelf, so inbound vessels sounded regularly to identify an approaching coast. When the lead weights of sounding devices were smeared with tallow or other sticky, soft material to pick up grains of bottom sediment, sounding also helped navigators fix their position as experience accumulated of the type of bottom found off various harbours and coasts.

The measurement used to define depth was the fathom. A fathom is about 1.8 m (6 ft), the wingspan of an adult male with arms outstretched. To sound the depth of a body of water, the navigator would heave the sounding lead overboard and watch the attached line run out until its progress slowed or stopped, signalling arrival at bottom. Then, starting from that point, he would haul in the line, one wingspan at a time, and count the number of fathoms of line that had been paid out to reach bottom. Even in relatively shallow waters, recognizing the moment the lead hit bottom required some skill, as did the technique of sounding while under way, which required throwing the lead forward of the vessel's bow so that the line was straight up and down for an accurate measurement by the time the vessel drew alongside.

The intimate connection between the human body and this unit of measure reflects visceral understanding that navigators had of the ocean, knowledge that was not available to ordinary sailors. In Rudyard Kipling's novel *Captains Courageous* (1897), Captain Disko Troop's fleet-wide reputation as the best fisherman and navigator rested in part on his ability to tell where his schooner was on the Grand Banks by feeling, smelling and tasting the bottom sediment retrieved by the sounding lead. One of his crew, Tom Platt, was exceptionally talented at deploying the lead, and the ability of the young protagonist, Harvey, to learn this skill signalled his rise within the ranks beyond ordinary sailor to future navigator.

Just as the working knowledge of navigators and sailors was instrumental to establishing ocean routes, so too did mariners' craft spark the discovery of the ocean's third dimension. Atlantic sailors who grew adept at sounding contributed this invaluable skill and habit to exploration of unknown waters. Whalers were also in the vanguard. Until early in the nineteenth century, most whaling was conducted from bases on shore. The preferred prey were 'right' whales, so-called for their inviting characteristics of swimming near shore, yielding plenty of oil and floating when dead. As whale populations near land declined, their hunters boarded larger vessels to venture farther afield. The introduction of the trywork, a shipboard furnace for rendering blubber aboard seagoing vessels, severed the link to land and sent whalers out to roam parts of the open ocean where traders and navigators, intent on moving between known places, never went.

American whaling captains, for example, were familiar with what became known as the Gulf Stream before men of science thought to ask. Benjamin Franklin, as Deputy Postmaster General for the American colonies, sought the answer to the puzzle of why mail-carrying ships sailing a northern route from England to the American colonies took weeks longer to cross than vessels plying a more southern route. His cousin, a Nantucket ship captain named Timothy Folger, knew the answer. American whalers who criss-crossed the Atlantic Ocean in all directions had observed that whales avoided certain areas, and noticed that the colour and temperature of these waters differed from surrounding ones. Merchant captains, by contrast, had no reason to notice these features. The detailed records kept by whalers made their working knowledge available when men of science began to enquire. Franklin used a sketch made by Folger as the basis for a published series of charts of the Gulf Stream, including a 1786 version that served as the basis for virtually all Gulf Stream charts until 1832.

In addition to simply traversing strange waters, whalers developed a preference for hunting a new species whose habits introduced them to the profound depths. Prompted partly by declining numbers of other whales, the shift to sperm whales was spurred as well by the superior quality of the oil produced from their blubber, which found demand as a lubricant for the machines of industrial factories. Sperm whales also have in their head cavities a waxy substance, called spermaceti, that produces candles with a brilliant, smokeless flame. Valuable ambergris, a substance found in their intestines, was used to fix perfumes. Like right whales, their carcasses float and they are slow enough for men rowing wooden whaleboats to overtake. Unlike right whales, they feed on giant squid that live at great depths, and can therefore dive very deep – and sometimes they did so when struck with a whaler's harpoon.

Fantastic stories circulated about how sperm whales could remain submerged for hours and about how astonishingly deep they could dive. Scientific experts believed that life could not exist in the sea below about 300 fathoms, yet whalers sometimes had to splice together several 200-fathom or longer lines to prevent the escape of a harpooned sperm whale that dove straight down. Such fish stories made their way to the eyes and ears of men of science through a handful of educated whalers and scientists who gleaned information from mariners. The British whaling captain William Scorseby contributed significantly to knowledge of whales and also of the Arctic seas in which he hunted them, publishing the respected volume *The Northern Whale Fishery* in 1820.

A whale killed in the Pacific had the harpoon of a vessel then working in the Atlantic embedded in its blubber. Such tales supported arguments for expeditions to search for the Northwest Passage and began to attract the attention of the planners of major national exploring expeditions. For instance, the organizers of both the u.s. Exploring Expedition (1838 to 1842) commanded by Charles Wilkes and the lesser-known North Pacific Exploring Expedition (1853 to 1856) both consulted whaling captains who were familiar with the Pacific waters and islands these expeditions would visit.

The captains suggested a study of the seabed to support the whaling industry, following the logic that fishers could use knowledge of bottom type to find rich fishing grounds. The American hydrographer and naval officer Matthew Fontaine Maury also tried to convince Lt John Rodgers to conduct a seabed investigation as part of the North Pacific Exploring Expedition. As Franklin had, Maury bridged the gulf between the working maritime world and the gentlemanly domain of science, taking seriously the knowledge won by whalers and other mariners and incorporating their sea stories into his series of *Explanations and Sailing Directions to Accompany the Wind and Current Charts* (eight editions, 1851 to 1859). Better known to posterity for his work on winds and currents as well as his role in promoting international cooperation in meteorology, Maury also used his position as head of the u.s. Naval Observatory to compile information about whale sightings and kills from logbooks, with the goal of creating useful tools much like his charts of winds and currents that were reputed to shorten sailing times for navigators who used them.

The impulse Maury exercised to compile information about the ocean and display it graphically for the convenience of users was not unique to him. He was inspired by the work of the Prussian explorer and naturalist Alexander von Humboldt in physical geography. Well known for his travels and writings about the continents of South America and Africa, Humboldt began his career fascinated with the ocean, its currents and other physical features, and the distribution of its flora and fauna in response to their environment. Many natural philosophers, stirred by Humboldt's vision of the interconnectedness of organisms and the physical world, set about collecting data across large areas, and the ocean seemed ideally suited for this approach. Alongside his prodigious work on winds, currents, meteorological observations and whale distribution that directly supported maritime industry, Maury also produced the 1855 book *The Physical Geography of the Sea*, reflecting his ambition to make a lasting scholarly contribution honouring Humboldt's influence as well as adding to practical knowledge.

THE MOST STRIKING AND NOVEL nineteenth-century use for the ocean depended as much on freedom of the seas as global trade did. If sperm whaling stimulated curiosity about the ocean's depths, submarine telegraphy galvanized action to fathom them, literally and figuratively. Early efforts to measure depth in the open ocean were rare and sporadic, such as Sir John Ross's attempt during his 1817–18 voyage to Baffin Bay and the u.s. Exploring Expedition under Charles Wilkes which tried sounding in deep Antarctic waters. In 1840 Ross's nephew, Sir James Clark Ross, conducted the deepest soundings to that time during his Arctic expedition. The prospect of telegraphy transformed deep-sea sounding from an occasional experiment into the responsibility of government hydrographers, and eventually into the routine work of cable company employees.

Oceanic surveying and charting flourished in lockstep with the burgeoning maritime commerce spurred by the Industrial Revolution. Most hydrographers naturally focused on shorelines, harbours and approaches to land from long-established sea routes. When steam began to revolutionize ocean travel, it placed demands for knowledge of the ocean based on new, direct transit routes that made sense for vessels no longer beholden to wind power. Maury's energetic efforts to use science to improve sea travel broadened in the late 1840s to include experimental deep-sea soundings, some of which focused on lanes that steamships had begun to ply. His motive appears to have combined his own scientific curiosity about the physical geography of the sea, inspired by Humboldt, with the rationale that, if he could find deep water in areas where sailors had reported shoals, he could remove incorrect notations from charts. Mariners, confident of the accuracy of up-to-date charts, could proceed at top speed in proven deep water.

After three years of intensive efforts by naval officers working under his direction, Maury created the first bathymetric chart of the North Atlantic basin in 1853. Based on about ninety soundings, it included shaded zones marking contours of one, two, three and four or more thousand fathoms. Most of the measurements were made using very simple technology and a technique based on traditional sounding. At first hydrographers simply tied package twine to a cannonball, tossed it over the side, and tracked how much twine ran out until they judged that the weight had reached bottom. Then they cut the line and sailed on. A clever innovation by a young lieutenant trained by Maury resulted in a sounding device that recovered a small sample of deep-sea sediment. This practice required the additional labour of recovering the device, but the reward was confirmation that the instrument had, indeed, struck the sea floor. In 1855 Maury made a new chart, recording about 189 soundings, three times the number from the original chart.

Maury's charts included a line of soundings along the favoured steamship route between Europe and North America, north of the belts of trade winds that sailing vessels had long relied upon. This line also fell near a similar route proposed for the Atlantic submarine telegraph. Promoters hoped to lay a cable from Ireland to Newfoundland, close to the great circle, or the shortest distance across the Atlantic. While it appears that Maury may not have been aware of the Atlantic cable project at the start of his deep-sea sounding work, the prospect of submarine telegraphy immediately accelerated efforts to study the ocean floor.

Maury's bathymetric chart, the first ocean-basin scale chart of its kind, caught the attention of mariners and men of science, but other images he created of the sea bottom reached ordinary people. His famous

'Telegraph Plateau' illustrated the ocean floor feature his hydrographers coincidentally discovered at exactly the place where entrepreneurs planned to lay the cable. This flat plain about 2,000 fathoms deep was located near the great circle, the shortest point between the landmasses of the Old and New Worlds. The drawing shows a jagged and forbidding sea floor south of the plateau, revealing the prevailing belief that the plateau was a providentially placed exception to the mostly rugged sea bottom. This and other images, such as a drawing of the shells of unthreatening diatoms and other microscopic creatures found in the soft mud of bottom samples, appeared on front pages of newspapers and illustrated weeklies, testifying to the appropriateness of the ocean as a home for telegraph cables.

Americans working under Maury and also at the U.S. Coast Survey were most active in deep-sea sounding through the early 1850s. The end of the Crimean War in 1854 freed British resources for deep-sea work. Thereafter the development of the technology for deep-ocean measurement happened on British vessels, reflecting that nation's greater involvement in the transatlantic telegraph cable laying efforts. The primary entrepreneur for the project over the decades it took to complete was the American Cyrus W. Field. A pair of attempts in 1857 and 1858 involved both American and British vessels and resulted in a cable that worked briefly until it failed for electrical reasons. Partly due to the American Civil War, almost a decade passed before a second series of attempts. In 1866, the largest ship ever built to that time, the 211-m (692-ft) SS *Great Eastern*, laid a successful cable, and also managed to locate and complete a cable that had broken the previous year.

On land, the Atlantic cable was hailed as the eighth wonder of the world and welcomed as a communications revolution that would ensure world peace. The work that went into sounding along its route and to reimagining the ocean floor as a safe place to put telegraph wires constituted a cultural discovery of the deep ocean that extended well beyond the desks of naval cartographers, entrepreneurs and engineers. Enthusiastic newspaper coverage about ocean telegraphy's failures and

Lithograph entitled *The Eighth Wonder of the World*, an allegorical scene in honour of the Atlantic cable, 1866.

successes and also popular narratives of cable-laying voyages found readers among Victorian families who may also have had sheet music on their pianos of 'The Atlantic Telegraph Polka' or 'The Ocean Telegraph March', celebrating the cable or its oceanic home in the depths. Or perhaps the lady of the house might have a bottle of 'Ocean Spray' perfume created in honour of Cyrus Field and advertised in *Harper's Weekly*.[2] Enthusiasts might have purchased one of the 50¢ souvenirs made from the 32 km (20 mi.) of leftover cable sold to Tiffany & Company jewellers. Umbrella handles, canes and watch fobs fashioned with sections of cable were also available. Field had a watch fob made for himself with a few precious grains of deep-sea sediment set in it.[3] The spectacular achievement of spanning the Atlantic with telegraph wires brought the bottom of the sea into the minds, lives and homes of ordinary people.

THE ATLANTIC CABLE may have revealed the ocean's depths, but for many people discovery of the sea began at the shore, where the new passion for beach holidays set the stage for dramatic growth of scientific interest in marine life. Beaches did not always hold the attraction they now do. Poor people gathered seaweed and scrounged for usable objects that washed up. Rumours abounded that wreckers deliberately lured ships aground with lights in order to salvage the flotsam that washed ashore from the wreckage. Before its cultural discovery, the seashore was associated with cannibals, mutineers and shipwreck victims. Daniel Defoe's Robinson Crusoe rarely ventured onto the beach, preferring to stay safely inland. Respectable people avoided the shore.

The rehabilitation of the coast involved both mind and body. Genteel young Europeans visited Holland on the Grand Tour starting in the mid-eighteenth century to witness scenes they knew from Dutch seascape paintings. In the throes of Romanticism, they sought the sublime generated by the extreme calm or the exquisite violence of the water. Romantic artists turned to the shore as an ideal place for reflection, where the correspondence between marine and psychological depths might lead to self-knowledge. Of at least equal attraction was the new-found salubrity of seawater and seaside air. The healthful and social appeal of inland spas evolved into a mania for bathing in the cold waters of northern European shores and breathing the sea air. Aristocrats ventured to the beach first but the social appeal of seaside resorts ensured that upper, then middle, classes followed. American discovery of the beach lagged the European by about a decade. Strictly regulated hydrotherapy promised cures for melancholy, anxiety and spleen. The bather endured the terror of the sea to overcome her maladies but, paradoxically and disguised by the therapeutic alibi, discovered bodily pleasures. By the 1840s, as railroads extended from cities to the shore, virtually everyone could afford at least the occasional day trip to the beach, establishing an association between urban Westerners and the beach that continues today.

Proper Victorians sought to balance the pleasures of new bodily sensations and the fashionable social scene with morality. In their eyes the beach served as a portal to learning about the wonders of nature. Seaside cliffs offered an easy three-dimensional glimpse into deep time for observers who had begun to grasp the new sense of the Earth's history proposed by geologists. Reflection on the endless waves and horizon allowed visitors to envision the long stretch of time and the forces that created and reshaped coasts, rocks and cliffs.

The waves tossed treasures from the dark depths at the feet of beachcombers who ventured off the boardwalks and balconies to touch the sand. Beautiful shells had long been prized for eighteenth-century cabinets of curiosities but the vogue for beach holidays expanded natural history to include all marine flora and fauna. Middle-class holiday-goers prowled the shoreline, finding seaweed, shells, marine creatures and their remains. To learn more, they consulted the numerous popular books on marine natural history that appeared, including works by the deeply religious naturalist and science popularizer Philip Henry Gosse, such as *The Ocean* and *A Naturalist's Rambles on the Devonshire Coast* (1844), which celebrated marine life as a sign of God's creation and featured colourful underwater scenes of living sea creatures. The Anglican priest and Cambridge professor Charles Kingsley, who embraced the idea of evolution, wrote a marine natural history book, *Glaucus: or, The Wonders of the Shore* (1855), before penning what became the delightful and popular children's novel *The Water Babies* (1863), about a young chimney sweep who becomes a water creature in order to evolve morally and socially. The American educator Elizabeth Cary Agassiz, wife of the famous zoologist Louis Agassiz and promoter of women's education in science, wrote *Seaside Studies in Natural-History* (1865) with her stepson Alexander, who later became a well-known ocean scientist and invented new ways to sample the depths.

Marine natural history offered an uplifting excuse for men, women and children alike to indulge in the pleasures of the beach, and the seaside holiday context drew women into science. Margaret Gatty, whose contributions to algology were recognized and appreciated by leaders of the field, began studying seaweeds during a stay at the coastal town of Hastings to recover from the birth of her seventh child and a subsequent bronchial ailment. Bored by her uncharacteristic inactivity, Gatty read William Harvey's *Phycologia Britannica* (1846 to 1851) and began wandering the beach to look at seaweeds for herself. She returned home with a passion for their study and began writing natural history, ultimately publishing the respected *British Seaweeds* in 1863. Gatty recruited her entire family for collecting, transforming holidays into searches for unusual specimens.

While Gatty and her family pursued marine natural history more seriously than most, they were joined on the strand by many middle-class families seeking morally appropriate leisure activities. Families could also enjoy the brand-new hobby of keeping marine animals alive back at home with the invention of what proved to be an enduringly popular instrument of science and entertainment, the aquarium. By the mid-1850s, London had two suppliers of live animals, as well as a public aquarium. Gosse followed his very popular marine natural history books with one titled *The Aquarium: An Unveiling of the Wonders of the Deep Sea*

The Fountain Aquarium, one of many popular means for keeping marine animals in the home during the height of the mid-19th-century aquarium craze.

(1854). Within two decades, Britain had almost a dozen public aquaria, and all major European cities boasted one. As in the case of beach holidays, American embrace of aquaria shortly followed European example. Gosse's son, Edmund, reported his father's chagrin when he recognized that an 'army' of natural history collectors had 'ravaged every corner' of England's rocky tidepools.[4]

Although beach strollers found the occasional zoological or botanical treasure washed up after a storm, serious scientific pursuit of marine natural history required boats. The geologist and zoologist Edward Forbes, who grew up in a maritime community on the Isle of Man, played a key role in promoting natural-history dredging within the British scientific community in the 1830s and 1840s. He arrived at the University of Edinburgh in 1831 to study medicine and joined a group of professors and students who partook eagerly in field excursions to collect material for natural history study. Forbes, familiar with oars and oyster dredges from his father's involvement with local fisheries, contributed his maritime experience to the group's dredging outings in hired fishing vessels and rowboats. One landlubber among his student peers was Charles Darwin, who took his new-found dredging experience with him aboard HMS *Beagle*. Another was George Johnston, who became a physician and an active promoter of marine zoology in Northumberland and founded the Berwickshire Naturalists' Club.

Natural history clubs fostered marine science by providing a meeting place for discussion and exchange, and also because they organized dredging excursions larger in scale than individual collectors could manage. As railroads reached the coasts and enabled wider social access to the seashore, vacationing dredgers discovered yachting as a way to collect marine creatures from deeper water.

The yachting tradition that fostered marine science was not the socially elevated races and lunch parties that characterized the Royal Yacht Squadron's Regatta Week at Cowes. Instead, cruising offered a genteel version of the maritime world that carried marine naturalists away from social centres to zoologically rich, often remote regions, for a kind of do-it-yourself exploration. Compared with hired fishing vessels, yachts were more physically comfortable for naturalists and provided them with more independence to choose dredging sites and experiment with collecting gear.

Yachting also permitted women to practise science at sea, because they were accepted as members in many of the natural history clubs and welcomed aboard yacht cruises that were as sociable as they were scientific. The 1871 Edinburgh meeting of the British Association for the Advancement of Science featured a dredging excursion for about sixty members including some women. Two years later the inland Birmingham Natural History Club organized a week-long visit to Teignmouth to dredge from the hired yacht *Ruby*. The participation of women added to the sociability of the holiday outing, as evidenced by the resolution of members to make the excursion annual, 'especially as ladies were now for the first time admissible as members'.[5]

Although oceanographers and most historians believe that the field of ocean sciences has been dominated by men from the start, a view from the decks of nineteenth-century vessels reveals that scientific study of the ocean began not only aboard hydrographic surveying ships and fishing vessels, but on rowboats and yachts. Because of the holiday context of beach-going, amateurs, including women and children, practised marine natural history alongside professional scientists. After science became a profession in the mid- to late nineteenth century, specialists in control of universities, museums and other important institutions began to dismiss the efforts of amateurs. Probably the main reason their participation in marine science has been underappreciated, though, relates to the place of the famous 1870s voyage of HMS *Challenger* in historical memory.

Long considered the foundational event for oceanography, the circumnavigation voyage by the *Challenger* (1872 to 1876) instead represents the culmination of scientific interest in the ocean from many sources. The success of the 1866 transatlantic cable proved the efficacy of submarine telegraphy and strongly motivated ocean-floor investigation and governmental willingness to fund it. Naturalists, who had begun collecting at the waterline and reached deeper into the sea from the decks of rowboats and then yachts, wanted to understand the distribution of marine creatures and hoped to determine whether life existed at the greatest depths. Based on dredging operations in the Mediterranean, Forbes concluded that life disappeared around 300 fathoms. After his untimely death in 1854 at age 39, the active community of naturalist-dredgers he had helped create began to sample more deeply and continued to find living creatures. The dramatic recovery of a failed submarine telegraph cable from the Mediterranean in 1860, raised from 1,000 fathoms encrusted with unfamiliar marine life, thrust the academic debate into the public spotlight.

A group of British naturalist-dredgers with strong ties to the Royal Society recognized the benefits of the built-in expertise of hydrographers and crews of naval surveying vessels accustomed to deep-sea sounding and lobbied for government help to deploy dredges in the ocean's depths with their help. At the request of the Royal Society, the Admiralty willingly provided vessels for a series of summer research cruises in the 1860s to test the feasibility of sampling beyond several hundred fathoms. The scientists aboard HMS *Lightning* (1868) and HMS *Porcupine* (1869 to 1870) found life everywhere they looked, down to more than 2,000 fathoms. The debate about whether or not life existed at great depths gave way to more multifaceted questions about the nature of life there. Observers noted with fascination that some life forms discovered at depths of several hundred fathoms were previously known only as fossils. In 1866, the Norwegian naturalist Georg Ossian Sars, dredging near Lofoten, discovered a crinoid at 300 fathoms. He and the naturalists he consulted were familiar with crinoids, also called sea lilies, in coastal areas, but Sars's find resembled instead fossil stalked crinoids. The British professor who became *Challenger*'s chief scientist travelled to Norway to see the prized specimens. This and subsequent similar catches raised the prospect that the ocean's depths might hide many so-called 'living fossils'.

That prospect appeared fulfilled in 1868 when Thomas Henry Huxley, known as 'Darwin's bulldog' for his vociferous defence of evolutionary theory but also an explorer who served as surgeon-naturalist aboard HMS *Rattlesnake*, announced the discovery, in stored deep-sea bottom samples, of a protoplasmic primitive organism that he believed to be a precursor to higher life forms. One important challenge to proponents of evolution was the unanswered question about how life began. The new-found creature was christened *Bathybius haeckelii* after the well-known German zoologist Ernst Haeckel, who had recently proposed a third biological kingdom, the Protista, alongside the established animal and vegetable kingdoms.

In this era when specimens from the depths became available for study and popular viewing, discoveries such as *Bathybius* attracted widespread public attention. In 1822 a preserved mermaid had made its way to London, only to be declared by naturalists a fraud created from pieces of an orang-utan, baboon and salmon. In the U.S., P. T. Barnum exhibited a 'Feejee mermaid' in 1842, really a dead baby monkey attached to a fish tail. Although Barnum declined an invitation to be

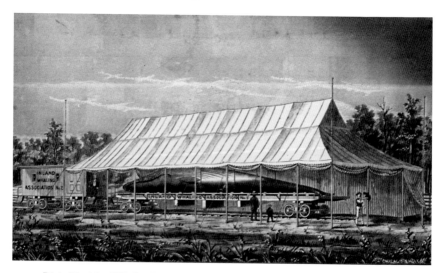

Print of the Inland Whaling Association's exhibit of an enormous blue whale, displayed in cities throughout the northeastern and midwestern United States in the early 1880s.

involved, entrepreneurs of what became known as the Inland Whaling Association exhibited a dead 18-m (60-ft) blue whale between 1880 and 1882, bringing the 'monster whale' by railroad flatbed car to northern cities on the East Coast and throughout the Midwest. Earlier a blue whale that washed ashore at Ostend, Belgium, in 1825 was de-fleshed and exhibited around Europe for seven years. In 1845, New York City residents could visit the 43-m (140-ft) skeleton of a sea serpent discovered and assembled by Albert Koch, a German immigrant, scientific collector and showman who had previously toured major American cities with a mastodon skeleton. The public greeted the monster enthusiastically, and the German King Frederick William IV purchased the skeleton, but experts pointed out that the bones came from six animals, not one. Despite this sea serpent deception, few sea monster specimens ever appeared for examination. Perhaps because of the plausibility lent by exhibits of fossil plesiosaurs and ichthyosaurs, sea serpents continued to hover in the realm of the feasible in the minds of at least some scientists.

Technological success at dredging and sounding in thousands of fathoms on the *Lightning* and *Porcupine* cruises seemed to promise answers to previously out-of-reach questions. Curiosity about oceanic life and, indeed, the origins of life itself inspired the voyage of the *Challenger*, which sailed from Portsmouth on 21 December 1872, and returned three and a half years, 127,580 km (68,890 nautical mi.), and 362 scientific stations later. The expedition's dredges, trawls and nets found life everywhere they sampled. Half of the 7,000 specimens retrieved proved new to science, and some came from depths over 3,000 fathoms. The results eventually appeared (it took 23 years) in fifty volumes that documented foraminifera, radiolaria, echinoderms, cetacean bones, medusa, copepods, crinoids and dozens of other groups of organisms. The mysterious *Bathybius* eluded *Challenger*'s scientists until the last months of the voyage, when it was discovered in preserved bottom samples. The suspicious chemist aboard investigated and debunked the celebrated proto-life form as a calcium sulphate precipitate formed when seawater reacted with the preserving fluid. Despite this disappointment, the *Challenger* collections and results formed the foundation for the emerging discipline of oceanography, whose origin owes much to the cultural discovery of the ocean.

IN PARALLEL WITH THE scientific and economic discoveries of the depths, the sea became an inexhaustible source of inspiration and promise for writers, including scientists and explorers alongside literary authors. Dramatically increasing literacy rates boosted readership in general, and the market for books about the ocean and ocean voyages greatly expanded in the nineteenth century, introducing more of the non-seafaring public to the world of ships and voyages. Maritime fiction helped transform the sea into the site of heroism and adventure and rendered ships into microcosms of society. Non-fiction accounts of voyages equally promoted this transformation, and those by scientists contributed to the refashioning of the ocean into knowable, usable and controllable space.

Literature about the sea appealed to a growing market of readers, including the first generations of Europeans and Americans to live in a world with regularly scheduled steam travel across the Atlantic. For the first time ever, people came into contact with the idea and the reality of the deep ocean in a way that had previously been limited to those whose families worked at sea, to residents of port towns or to people whose lives accidentally intersected with the ocean. Between 1815 and 1930 the greatest episode of human migration exposed fifty million Europeans to an ocean voyage. As overseas immigration peaked in the 1850s, the U.S. became home to a population who had experienced sea travel in mass numbers. The crossing experience deeply influenced immigrants, who carried nautical metaphors with them as they moved westwards. Most never sailed again, but the accomplishment of ocean crossing provided Americans with a shared historical past and perhaps fanned their taste for maritime literature. Fashionable steam travel transformed the Atlantic crossing from a life-changing emigration experience or a work-related necessity into a part of the social season for a wealthy and, increasingly, middle-class families. Both the u.s. and Britain embraced their identities as maritime nations early in the nineteenth century, but awareness of the sea particularly pervaded Britain, the island centre of a global empire.

American and British readers appreciated the new maritime novels. After Defoe's *Robinson Crusoe* (1719), a hiatus interrupted the production of popular maritime fiction until early in the nineteenth century, although the British writer Tobias Smollett employed naval characters and settings in satirical works such as *The Adventures of Roderick Random* (1748). Sir Walter Scott's *The Pirate* (1822) reignited the genre, and James Fenimore Cooper paid homage to Scott and his main character in *The Pilot* (1823), a tale about John Paul Jones's privateering along the Scottish coasts. Cooper, and also Washington Irving, America's first two successful professional writers, both wrote well-received sea stories. A frequent transatlantic traveller, Irving wrote stories of voyages, shipwrecks and pirate treasure, and also authored the widely read *A History of the Life and Voyages of Christopher Columbus* (1828), which blended fiction with fact and originated the myth that Europeans believed the Earth to be flat. Cooper employed his experience sailing as crew on merchant vessels and his subsequent naval service to portray sea work and life authentically. Writing after long voyages had become safer and more predictable, and at a time when most of the world's coasts and oceans had been explored, he presented the sailor's craft of routine maritime work as heroic and adventurous. Cooper's work travelled across the Atlantic, winning critical appreciation and inspiring sea fiction in both Britain and France. Perhaps it inspired Frederick Marryat, the British Royal Navy officer and literary acquaintance of Charles Dickens who published the semi-autobiographical *Mr Midshipman Easy* (1836) and other maritime stories.

As Cooper and Marryat had, many maritime novelists had experience at sea. Americans Richard Henry Dana Jr and Herman Melville numbered among the first professional writers after Cooper to base their accounts upon personal experience of working at sea as common sailors rather than recreation or leisure travel. Before this time, although retired mariners such as Marryat had turned into writers, educated landlubber writers would not have eagerly faced the dangers associated with working at sea. Some writers gained maritime experience aboard yachts, as had the English Romantic poet George Gordon, Lord Byron. French writers Jules Verne and Victor Hugo were also enthusiastic yachtsmen. Verne, who wrote most of his imaginative tale *20,000 Leagues under the Sea* (1870) while cruising on his yacht *Saint-Michel*, also travelled across the Atlantic as a passenger on the *Great Eastern* the year after it had been used to lay the successful 1866 transatlantic cable.

The new breed of maritime writers were often well-educated young men who chose to go to sea yet remained conscious of their social distance from the sailors who became their workmates. Dana, a well-heeled Harvard student in the early 1830s, went to sea for his health, embracing current convictions about the salubrity of sea air. Eschewing a fashionable European grand tour he instead enlisted as a merchant seaman, publishing *Two Years before the Mast* in 1840 as he embarked on a law career in which he advocated on behalf of common sailors and in favour of the abolition of slavery. While he hoped his book would educate the American public about the brutality and ugliness of maritime work, his tale also inspired many landlocked boys to ship out. Melville, who went to sea when his merchant father's death left the family in financial straits, admired Dana's description of rounding the infamous Cape Horn, which, as he put it in his own novel *White Jacket* (1850), 'must have been written with an icicle'.[6] Melville's experiences on merchant and naval vessels and his time in the Marquesas Islands provided grist for many of his books including *Typee* (1846) and *White-Jacket*, among others, while his service on the whaleship *Acushnet* inspired *Moby-Dick* (1851). As Melville's eager reading of Dana's narrative makes clear, the educated young men who set sail did so having read voyaging accounts and often with the expectation of writing about their own experiences.

The first scientists to make the ocean itself a focus of study went to sea in the thrall of sea fiction that presented voyaging as a route to adventure. They also read narratives written by leaders of famous exploring expeditions such as Charles Darwin, Alexander von Humboldt, Captain James Cook and any explorer who had ventured to the particular corner of the globe to which they were headed. Scientists who accompanied the circumnavigation of the *Challenger* carried the memory of Robinson Crusoe's adventures with them. As they approached the island of Juan Fernández, several wrote in their personal journals about Alexander Selkirk, the real castaway whose experiences there inspired Defoe's tale. En route to St Paul's Rocks, 800 km (almost 500 mi.) off Brazil, they consulted Darwin's account of the 1831 to 1836 *Beagle* voyage about his activities there and prepared the bait and hooks he recommended for shark fishing. Most of the *Challenger* scientists published popular accounts of the expedition. Several of the naval lieutenants also did so, and one member of the crew, the steward's assistant, sent letters home that he intended to copy into a single volume for his family after the expedition. The literary context of voyaging made scientific adventurers think of their own seagoing as part of a tradition of writing and publishing. Along with their literary and labouring counterparts, scientists set sail in pursuit of a personal experience, anticipating the opportunity, danger, heroism and self-transformation attendant on the new nineteenth-century encounter with the ocean.

Writers of sea fiction read the books scientists wrote about the ocean, incorporating into their novels state-of-the-art knowledge about the sea. Edgar Allan Poe's *The Narrative of Arthur Gordon Pym of Nantucket* (1838) bears the influence of the so-called 'Hollow Earth' theory that inspired the polar investigations undertaken by the u.s. Exploring Expedition of 1838 to 1842 and also the many Arctic expeditions of the 1850s that tried to find Sir John Franklin, or remnants of his lost expedition, by seeking the reported open polar sea. Through his work as an editor, Poe was well aware of public interest in both sea literature and scientific discoveries about the sea. Melville has Ishmael refer to William Scorseby and also Thomas Beale, author of *The Natural History of the Sperm Whale* (1839), in the 'Cetology' chapter of *Moby-Dick*. After drafting the chapter 'The Chart', in which Captain Ahab retreats to his cabin after a storm to pore over charts and mark them with information gleaned from old logbooks, Melville learned about Maury's preliminary sketch for a global whale chart, a project which Melville employed to lend credence to the hunt for a single whale in the vastness of the sea. Victor Hugo likewise employed the new ocean science as he wrote *Travaillieurs de la mer* (Workers of the Sea, 1866), including work by Maury, as well as Jules Michelet's *La Mer* (1861).

Visual artists also drew from scientific works to evoke or depict the ocean's depths. In 1864, the American artist Elihu Vedder achieved his first public success exhibiting his oil painting *The Lair of the Sea Serpent*. Vedder, whose work was admired by Herman Melville, sketched studies of an eel as the model for his sea monster but took inspiration also from the fantastical works of Gustave Doré, illustrator by that time of works by Byron but who went on to create illustrations including strange sea creatures for Hugo's *Workers of the Sea* and the 1870 edition of Samuel Taylor Coleridge's *Rime of the Ancient Mariner*. The apparent serenity of

Elihu Vedder, *The Lair of the Sea Serpent*, 1864, a sinister painting which created a stir in the art world when first exhibited. In 1889 Vedder painted another work with the same name.

An undersea version of a Hudson River school landscape painting, Edward Moran's *Valley in the Sea* (1862), may have been inspired by the first Atlantic telegraph cable.

Vedder's scene, a dune overlooking the beach and sea beyond, is unsettled by the open eye and utter stillness of the creature lying in wait and evoking the terrifying mystery of the ocean's depths. Edward Moran's 1862 oil painting *Valley in the Sea* actually depicts the undersea realm in a panorama reminiscent of the Hudson River landscape style of American art. Probably commissioned by its first owner, James M. Sommerville, a Philadelphia physician, amateur artist and naturalist-dredger, the painting was likely inspired by the 1858 Atlantic cable and appears to represent visually the broad, flat oceanic valley described in Maury's 1855 *Physical Geography of the Sea*. Sommerville, who in 1859 published a small scientific book titled *Ocean Life*, also tried to represent the sea floor artistically himself, collaborating with another artist to produce a watercolour similarly titled *Ocean Life*. The colourful underwater scene, crowded with many of the species discussed in the book, was used to make lithographic reproductions to illustrate it.

Maury's work, which was threaded through so many literary and artistic representations of the ocean's depths, influenced Jules Verne, who wrote *20,000 Leagues under the Sea* (1870) with a copy of Maury's *Physical Geography of the Sea* beside him. Verne guided the *Nautilus* around the world's oceans along a route that shadowed Maury's discussion of geographic regions of the ocean. Whole passages of Verne's *20,000 Leagues under the Sea* mirror Maury's text, a common enough feature in the world of nineteenth-century publishing that demonstrates the importance Verne placed on employing up-to-date geography and science in his work. In addition to incorporating his own sailing experience and Maury's text, Verne questioned the *Great Eastern* crew about the cable-laying work. He was also inspired by a visit to the 1867 World's Fair in Paris, which featured an aquarium and daily demonstrations of new diving equipment. Verne's oceans were equally fact and fancy, just as the mid-century discovery of the depths was as imaginative as it was scientific, as intellectual as it was technological and as personal as it was official.

MID-CENTURY DISCOVERY of the ocean's depths and remote blue waters proceeded through museum visits, aquarium keeping, maritime novel reading and amateur seaweed collecting as much as through government hydrography, professional marine zoology or submarine cable laying. The home was as important as natural history workrooms or ships. Returning seaside vacationers proudly exhibited aquaria or shells in Victorian drawing rooms alongside pianos displaying sheet music celebrating mermaids or the Atlantic cable. People from all walks of life could read in newspapers and popular magazines about the first transatlantic yacht race in 1866 that became the international competition still known as the America's Cup. Parents dressed their children in sailor suits inspired by the fancy nautical costumes favoured by yachtsmen, famously including Queen Victoria's husband and son. By the end of the century, the sailor suit had become a widely popular style for both boys and girls, and even for women. Hobbies, clothing, collections, reading materials, and even restaurant menus featuring choices such as 'deep sea flounder', reflected a spreading acquaintance with the ocean, including its vast third dimension.[7]

Familiarity with the ocean and things maritime reflected a radical new posture towards the ocean that emerged in the nineteenth century. For the first time in history, the sea became a destination. Before, ocean travellers had set sail for other lands, using the ocean as a byway. Naval fleets sailed in search of enemy ships to fight. Navigators followed proven routes. Explorers sought new coasts, direct sea routes and safe harbours. Fishers sought fish or whales and, even when whalers first pointed their bows toward the open ocean in pursuit of sperm whales, their goal was their prey. The point of setting sail was always to get back to land expeditiously. The shift was a social one, wrought by the whalers, novelists, scientists and other seafarers who began to embrace the notion of going *to* sea. Many professional mariners did not share this essentially land-lubberly perspective, but the new seafarers who embarked for the purpose of experiencing the sea understood the ocean itself as their destination. Their stories conveyed to armchair sailors the sense of the sea as a place where people tested themselves against nature, creating a stage for heroism, recreation, personal transformation, national triumph or control of natural forces. Sounding, science and submarine telegraphy rendered the sea culturally visible and helped enlarge the human relationship with the ocean to encompass its entirety.

FIVE

Industrial Ocean

Don't haul on the rope, don't climb up the mast
If you see a sailing ship, it might be your last.
Just get your civvies ready for another run ashore
A sailor ain't a sailor, ain't a sailor anymore.

Well they gave us an engine that first went up and down
Then with more technology the engine went around.
We know our steam and diesel but what's a mainyard for?
A stoker ain't a stoker with a shovel anymore.

– Tom Lewis, 'The Last Shanty', in *Surfacing* (1987)

THE MID-NINETEENTH-CENTURY cultural discovery of the sea added rapid communication, recreation and science to the human uses for the ocean. Steam and iron opened continental interiors to capitalist development, but the effects of industrialization likewise extended into the sea. Traditional uses of the ocean, especially fishing, shipping and the extension of power, intensified dramatically in the twentieth century, tightening the connections between the ocean and especially Westerners but others as well. Electricity, along with the steam that replaced wind power, quickened the tempo of trade and travel, shrinking the effect of global distances. Submarine warfare ensured that the sea remained an important site for the exercise of geopolitics during peace as well as war. Flowing against these currents, the number of people involved with maritime work declined, and the work that remained became increasingly invisible to the general public. Cultural perception of the sea shifted in response, from viewing the ocean as a workplace to considering it a site of reflection and renewal, a place outside time offering retreat from the modern world.

Cod, harvested in a traditional Norwegian fishery at Lofoten,
drying on racks ashore in 1913.

WHILE CHANGES ASSOCIATED with the Industrial Revolution tightened the connections between people in modernizing nations and food or other resources from the sea, sea fisheries had been exploited intensively long before that time. Fishers of various cultures and times understood the ocean as three-dimensional long before the Atlantic cable and the aquarium drew political and personal attention below the waves. The Atlantic cod, *Gadus morhua*, resides near the sea bottom in cold water throughout the North Atlantic and has been an important commodity in international trade since the Vikings began drying it for use as a food source for their long voyages. Unlike most fisheries before the twentieth century, cod grounds lay far from population centres for much of the millennium they have been fished. The distance encouraged the salting of cod, whose flesh is well suited for this preservation technique. Cod fishers who worked on the Grand Banks off Newfoundland gained intimate knowledge of the sea-floor and surface conditions, which permitted them to move around, even in fogs and during storms, to find the fish they sought. Rudyard Kipling's character Captain Disko Troop 'was, in fact, for an hour a cod himself, and looked remarkably like one', as he considered the ocean environment from the perspective of a 20-lb cod in order to decide where to fish next.[1]

Abundant cod became the symbol for Massachusetts and the underpinning of the economies of New England and Atlantic Canada. North American farmers fished part-time until the nineteenth century, at which time both agriculture and fishing transformed into commercial enterprises, resulting in separate systems for exploiting resources of land and sea. Other than cod, colonial fisheries concentrated initially on rivers, much as European fisheries had until freshwater species there were exhausted in about the eleventh century. In northwestern Atlantic coastal communities, river fisheries provided resources that remained in the community, while sea fisheries yielded exports.

River-, inshore- and shellfisheries were regulated locally through the mid-nineteenth century, in part to clarify who had access to the resource. Declining catches have been a familiar part of fisheries where- and whenever there are written records. Perceptions of overfishing, often articulated by fishers using traditional gear who feared competition from new, more intensive harvesting techniques, inspired calls for government regulation. So did concerns about mills or the pollution of rivers and estuaries from industrial and human waste affecting oyster and salmon fisheries. For sea fisheries, the embrace of freedom of the seas made it inconceivable to consider restricting them. Sea fisheries' traditional explanations for lowered catches blamed local overfishing and fish moving to new locations, so seeking new grounds became the usual solution, a strategy reminiscent of the terrestrial pattern of clearing new fields for agriculture. As with trade, nations with the knowledge and the power to use the sea promoted sea fishing. Because fishing supported national economies and trained sailors who could be conscripted in times of war, governments employed bounties and treaties to encourage it.

Although mechanization is often associated with problems created by industrialization, fishing and other marine resource exploitation could have profound effects on species and ecosystems in the absence of new harvesting technology. The otter and fur seal fisheries in the Pacific, for example, grew in the early nineteenth century in step with the ambitions of Russia, Spain and the u.s. to extend power around and over that great ocean. Vast Russian harvests were often made by enslaved natives from the Aleutian Islands using traditional gear and methods. The push behind the meteoric growth of the whaling industry after the War of 1812 included both the paucity of prey for whalers in the Atlantic, who were forced to sail far from home ports to search for whales in the Pacific, and the high demand for lubricants for industrial machinery. This industry used sails, wooden vessels, hempen lines, iron harpoon points and human labour to pursue, kill, dissect and render the blubber of sperm, humpback and other whales.

The North Atlantic halibut fishery, which emerged suddenly in the 1840s, exemplifies the dramatic economic, social and environmental effects that non-mechanized fisheries could exert. Halibut dwell in coastal waters and on offshore grounds that are also the preferred habitat of cod. Initially discarded as by-catch, halibut transformed into a marketable commodity, in demand by immigrants from Catholic countries who ate fish on fasting days. Other eager consumers included urbanites, whose tastes shifted from salted or pickled fish, ideally with a somewhat glutinous texture, to fresh fish with firmer flesh. Prosecuted initially by individual fishers with hook and line, the cod and halibut fishery multiplied the harvest in the late 1840s by adopting trawl-lining, a technique that involved two fishers in a dory laying hundreds of feet of line to which baited hooks were attached about every 2 m (6 ft). In Kipling's novel *Captains Courageous* (1897), the crew of the *We're Here* fished at times with hand-held lines but also used the tub-trawling method. In a decade, halibut were fished out of Massachusetts Bay, then subsequently from the Gulf of St Lawrence and Georges Bank, and then from deeper water. Without mechanization, the fishery rendered Atlantic halibut commercially extinct by the 1880s.

Trawl-lining, which boosted the number of hooks per fisherman from four to four hundred, was joined by other new harvesting strategies and techniques that increased catches in the second half of the nineteenth century. Weirs, while not new, found more intensive use, as did net fisheries generally after the 1865 invention of machines to make netting. Weirs and pound nets, which are standing gear that direct fish travelling downstream or along a coast, entrap them where they can be collected with scoop nets or other means. Purse seines, nets set to encircle a school of fish, could be drawn closed at the bottom, allowing the capture of entire schools. Gill nets, also an aboriginal technology like weirs, were adopted to capture salmon, destined for industrial canneries, before they entered their natal rivers. Each of these technologies, even before mechanization of fishing vessels or of the gear itself, initially increased catches but also raised cries of concern about overfishing from fishers using traditional gear.

Industrialization, if not mechanization of fish harvesting, exerted a significant effect on halibut and other fish populations via the ice industry and, later, refrigeration, as well as railroads. Canning and other processing innovations that were developed in the 1880s multiplied the number of species that could profitably be landed and distributed for sale farther from the docks. Monterey's Cannery Row, made famous by John Steinbeck in his 1945 novel, relied on plentiful sardines until stocks crashed in the 1950s. Canneries along the coast of northern California, Canada and Alaska transformed salmon from a regional product to a global commodity. Starting in 1903, when a southern California packing plant canned albacore when local sardines failed, canned tuna grew to become a staple food in the u.s., Europe and Japan. Each successful new seafood product involved not just technology, but equally required marketing and swaying of consumer tastes.

In addition to the harvest of more and a wider variety of food fishes, fisheries developed so as to catch the small fish that fed the bigger fish, transforming them into oil, fertilizer and later animal feed. Starting on a modest scale early in the nineteenth century, menhaden were caught in New England waters and used for bait or pressed for oil. Also known as pogy or bunker, menhaden were at their fattest after migrating north

Koenig & Sons Cannery, on the US Pacific coast, showing canned salmon.

while growing to full size, yielding at this point in their life cycle the maximum amount of oil. The fishery adopted first the purse seine and later coal-burning steam engines to power vessels. It expanded south to follow the fish as stocks declined in northern waters, never to return in commercial numbers, and disappeared entirely from the Gulf of Maine. Processing factories shifted from oil, used for making soap and paint as well as for tanning and curing leather, to fertilizer production, for which the thinner southern fish were better suited. The emergence of the petroleum industry and the decline of the guano trade from the Pacific encouraged this swing. Despite falling catches, many observers at the time believed that migration rather than overfishing caused the decreased yields.

Worries about overfishing cropped up frequently in the second half of the nineteenth century focusing on various harvesting techniques, gear or industrial practices said to be harming fish stocks. One particular technology galvanized concern, even as its efficiency guaranteed its embrace by the modernizing fishing industry. Trawling dated from the fourteenth century but was rarely used before its limited adoption in England and the Netherlands in the seventeenth century. Pulled by sailing vessels, trawls targeted bottom fish such as cod, ling, hake, whiting, sole and plaice. Starting in the mid-nineteenth century, the experimental application of steam power to trawling took place in England and Scotland, resulting in the establishment of a steam trawl fishery from the 1880s onwards. 'Otter boards', an 1892 innovation that drastically enhanced the efficiency of trawling by about 35 per cent, were attached between each trawl line and the mouth of the net and functioned as big barn doors to keep the net's mouth wide open as it dragged along the bottom.

Otter trawls pulled by steam vessels radically altered the relationship between fisher and fish, according to historian Jeffrey Bolster. Earlier fishing technologies tended to target particular species, so that their operators mostly caught the kind of fish they sought. Also, traditional fishing gear was passive, whereas otter trawls represented active pursuit of fish. Finally, in contrast to gear such as hook and line or gill nets, which caught individuals of a particular size, otter trawling caught anything and everything on the bottom, including young fish and market-sized ones, as well as whatever else grew or lived on the sea floor. Drawing upon an analogy to ploughing the Earth, many contemporaries believed that trawling's effect of disturbing the bottom likely increased the productivity of the sea. Traditional fishers, however, thought that the action of trawls on the sea floor killed spawn and juvenile fish. Like earlier conflicts over fishing gear, those that arose over trawling pitted against each other opposing groups of fishers who each claimed direct knowledge based on their work at sea.

THE MECHANIZATION OF trawl fishery enabled its development on an industrial scale and its geographic extension over vast areas of sea floor. Trawling spread throughout western Europe, as well as to

Japan and North America, but its early and intensive use in Europe has made the North Sea ground zero for our efforts to understand the effects of industrialized fishing. While sailing smacks ranged about 160 km (100 mi.), steamers could fish to 640 km (400 mi.) away from port and in deeper waters, which they were compelled to do as areas closer to home began to yield smaller catches. On average, steamer catches were six to eight times larger than those of sailing vessels, creating an initial bonanza in a newly fished area but ever more quickly leading to reduced yields and the need to seek new grounds.

The illustration below shows the trajectory of English trawling from 1833, intially occurring only in the small area north of the Strait of Dover between England and the Netherlands, spreading by 1845 along the coast north to the Scottish border and west towards the Dutch and German territorial waters. Twenty years later, the deeper waters in the middle of the North Sea were fished, and thirty years later, the areas

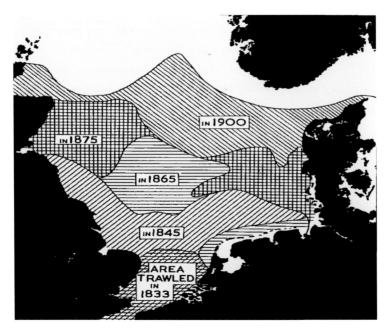

Diagram illustrating the spread of English trawling throughout the North Sea during the 19th century.

off Scotland and Denmark. By late in the nineteenth century, the entire British trawler fleet consisted of steamers pulling otter trawls that by 1900 had pushed north towards Norway, fishing every part of the North Sea. As was also the case for the Atlantic halibut fishery, railroads and the ice industry fuelled the expansion of fisheries as demand from growing and urbanizing populations made the flow of food fish a matter of government concern throughout Europe.

The expansion of trawling in England proceeded in step with calls by traditional fishers for its restriction, based on worries about destruction of young fish but also reflecting competition with older fishing methods. In response to persistent complaints, the British government convened a series of enquiries between 1860 and 1883 to investigate the alleged stock declines caused by trawling and to consider whether existing or new regulations were warranted. These inquiries reflected the trend of governments turning to experts to provide scientific and technical advice, in fisheries and in other sectors, to serve as the basis for solving novel problems, such as the so-called 'trawling question'. Inspectors for oyster and salmon fisheries in many northern European nations formed the nucleus for an emerging civil service dedicated to fisheries that worked in concert with fisheries science to place management on a scientific basis.

In England a definitive and influential answer to the 'trawling question' materialized from results of government enquiries and the strength of laissez-faire ideology. An 1863 investigation concluded that existing regulations, based on insufficient knowledge, should be removed. An 1883 report found statistics inadequate to demonstrate declines of fish stocks despite the pattern of trawlers having to move farther offshore to maintain catches. The 1883 report also exonerated trawling of the charge of destroying spawn, based on new findings that fish eggs float. At the International Fisheries Exposition the same year, the respected scientist well known for defending Darwin's evolutionary theory, Thomas Henry Huxley, issued his judgement that sea fisheries were inexhaustible, a statement that profoundly influenced fisheries management in many countries. His cautionary coda – that his assessment was based on then-common gear and practices – went unheeded. Appointed in 1880 as a fishery inspector, Huxley devoted minimal effort to that role and mainly used his position to question the propriety of government intervention, except perhaps to promote the fishing industry to feed growing populations.

In a number of northern Atlantic countries, including Canada and Norway, the state became the main facilitator of modernizing the fishing industry, though more of the major fishing nations left that task to private capital. Vertical integration of the fishing industry paralleled the same trend in the steel industry, with companies buying vessels, hiring fishers as wage labourers and also investing in net and twine companies, wharves, processing facilities and distribution operations. Trawling was routinely favoured by government managers over other forms of capture because of its efficiency and also because of the ease for them of working with fewer big businesses rather than numerous independent fishing operators.

Efficiency became the rallying cry for governments embracing science as a tool to maximize natural resource use. Forests in Germany served as the nursery ground for what emerged as European and American understanding of conservation. While conservation included the goal of ensuring resources for future use, scientific forestry introduced the concept of optimal sustained yield, which confidently expected maximal harvests year after year so long as a sufficient standing stock of trees remained. Applied to fish, the 'Gospel of Efficiency' logic suggested that not catching fish was wasteful.[2] Science could maximize both catches and profits for the industry, in theory addressing overfishing as well as the persistent poverty of the fisheries sector.

In reality, the ideals of what became the progressive movement fostered centralization of the fishing industry, modernization of harvesting and distribution of fish, and the exertion of federal government control over fisheries policy, formerly made at the local level. Fisheries managers embraced the relatively new tool of statistics to gain control

Photograph (1908) showing a catch from experimental fishing on the U.S. Fish
Commission vessel *Albatross*, sent to explore new grounds and find new species
for fisheries to exploit.

over the industry. Equally useful for science and administration, statistics
gave fisheries managers knowledge of catches which could, for example,
be used to anticipate processing and distribution needs.

Nations which aimed to control fisheries through accurate prediction of future catches invested in scientific research. Twenty-six marine
laboratories were founded throughout northern Europe and in the U.S.
and Canada by 1910. Many were created in appreciative imitation of the
Naples zoological station and, like it, contributed to the development
of experimental biology, but most also conducted fisheries research,
which formed the *raison d'être* for their funding. Some station scientists and directors also held government positions as fisheries officials,
such as Paulus C. Hoek in the Netherlands, Johan Hjort in Norway
and Edward E. Prince in Canada, exemplifying the tight connections
between science and emerging fisheries policy.

By the turn of the century, a group of marine scientists from Scandinavia and other northern European nations had concluded that effective research on fish, and on ocean currents as well, required international cooperation. Governments concurred and supported a new intergovernmental body established in 1902, the International Council for the Exploration of the Seas (ICES), which conducted international investigations and advised governments about fisheries matters. Its Overfishing Committee, for instance, analysed fisheries statistics from 1870 onwards and discovered that steam trawling had reduced catches relative to fishing effort. Fisheries scientists recognized the real possibility of overfishing bounded populations, such as North Sea herring or plaice, but insisted that there would always be new populations in the vast ocean to exploit. As a result, fisheries science focused first on positive measures to improve catches, such as hatcheries, transplantation of young fish to new grounds or locating new stocks to exploit.

Fisheries marched further offshore with mechanization, larger boats, and a more efficient, vertically integrated industry. Diesel and gasoline replaced steam. Europeans fished off Iceland by the 1890s and in the southern Barents Sea and White Sea early in the new century. In most fisheries, the number of fishers increased, as did the size and value of boats and gear, but catches dropped. American catches declined by 10 per cent between 1880 and 1908, for example. European fisheries scientists christened the First World War 'the Great Fishing Experiment', hoping to study the resumption of fishing after the war to understand how fishing affects stocks. Their efforts to protect fish stocks that recovered during the war, through increased mesh size or other gear regulations, failed, and fisheries boomed as more efficient gear and practices yielded catches larger than those of the pre-war years.

To many, the dramatic increase in scale of yields bolstered the optimistic belief of essentially unlimited marine resources. Fishing industrialized, transformed not only by new technologies but – and powerfully – by the idea of rational management to maximize fish yields. The new field of fisheries science set the tone for the massive-scale exploitation of living marine resources that proceeded in the second half of the twentieth century, including fisheries for fishes intended for food and small fish destined for reduction, as well as for the great whales.

WHALING, TRADITIONALLY CONSIDERED fishery by whalers despite its mammalian prey, followed a similar trajectory of industrialization, mechanization and scientific management. Nineteenth-century whalers sought blubber to render into oil and, later in the century, baleen for use in corsets, umbrellas and other products that called for its stiff and strong yet flexible qualities. Even before the discovery of petroleum, whaling was in decline until technological developments allowed the capture of faster-swimming rorqual whales, such as blues and fins, which had escaped hunters of the wooden age. Perhaps the critical innovation in modern whaling was the use of whales for food, especially by the margarine industry and particularly after the Second World War to feed hungry populations in Europe and Japan.

The modern whaling industry swept aside wooden whale ships and hand-thrown harpoons in favour of exploding harpoon cannons mounted in the bows of fast, steam-powered catcher boats that delivered carcasses to shore stations and, later, massive factory ships. With deadly effectiveness, whaling technology facilitated an almost completely unregulated, multinational fishery centred in the Southern Ocean around Antarctica. Even the ocean's giants, the blue whales, could be hauled up enormous slipways and processed in order to extract oil for margarine, or for ingredients for lipstick, shoe polish or explosive nitroglycerine; meat for animal or human consumption; and vitamin A from whale livers. The value of the meat and oil products from a single large whale, the most valuable of wild animals on the planet, could be as high as U.S. $30,000 in 1960 (worth about $240,000 today), although historians who recognize how heavily whaling was subsidized question the actual value of products derived from whales if expenses were deducted.

Through the 1920s, whales seemed plentiful in the Southern Ocean, and whalers sometimes used the best parts of carcasses, discarding the rest. The progressive conservationist legacy inspired some commentators to call for full utilization of whale bodies and also the protection of whale stocks. Roy Chapman Andrews, a curator at the American Museum of Natural History, recounted his desire to take up the harpoon and try killing a whale himself. During a 1909 whaling voyage he accompanied to study whales of the North Pacific, he convinced a Norwegian gunner to let him shoot one. Andrews's fantasy of whaling as big-game hunting emerged from the activities of the elite and urban hunters of the Boone and Crockett Club who criss-crossed the terrestrial wilderness of North America bagging trophies and, when back at home, advocated for the preservation of game animals.

In whaling nations, scientists and conservationists such as Andrews began to press for conservation measures to protect whale populations, including what seemed to them common-sense steps such as banning the killing of rare right whales and of the females of any species that were pregnant or with suckling calves. Yet gunners could not always tell if an animal in their sights was male or female, much less pregnant. While whalers knew how to find and catch whales, much was unknown about these creatures that mostly remained hidden below the surface.

Naturalists took advantage of the twentieth-century whaling industry to study whales, donning hip boots to measure and dissect carcasses as they were being processed at shore stations and on greasy slipways. Many of them drew parallels between the ocean giants and the bison of the American West, articulating the strong sense of the inevitability of their commercial loss. In 1910, Frederic A. Lucas, director of the Brooklyn Museum of Natural Science, insisted the time had come to rescue the bison's 'warm-blooded cousins from the ocean plains'.[3] In 1940 the world's foremost cetacean biology expert, Remington Kellogg, alerted readers of his *National Geographic* article, which introduced all the known species of whales, that these animals were 'heading toward the same fate that pursued the once-vast herds of American Buffalo'.[4]

The association of the ocean with wilderness akin to the American West, and its loss, formed part of a wider shift in cultural perception of the ocean.

AGAINST THE BACKDROP of increasing industrial use of the ocean came the contradictory trend of the decreasing visibility of such work. Between the sixteenth and nineteenth centuries, maritime activity supported nations and expanding empires. Coastal peoples overwhelmingly worked in occupations related to the sea and seafaring, whether directly or in supporting industries, part-time or year-round. Towards the end of the nineteenth century, maritime work declined relative to land-based work. Capitalist development turned to the interiors of continents, often fuelled by the profits from whaling, shipping and other maritime activities that were invested inland. At sea, efficiencies from mechanization reduced the number of people employed in maritime work.

Industrial whaling in the Ross Sea in 1922: three Norwegian workers from the FF *Sir James Clark Ross* flense the whales in calm weather.

By the end of the century, many coastal fisheries were in decline. Offshore fisheries integrated vertically into large corporations that were often not primarily maritime in orientation, further eroding the sea's visibility. Shipping companies turned to steamers, which employed fewer sailors and increasingly insulated passengers from the sea. As the tempo and reach of shipping and other maritime activities increased, the decks of merchant vessels, and also of whaling ships and fishing craft, began to take on characteristics mirroring terrestrial industrial work sites, with longer work days, fewer workers and lower pay. Companies that continued to use sail technology cut crews to the bare minimum, prompting an efflorescence of shanty singing because, 'as sailors say a song is as good as ten men', or so Richard Henry Dana explained in *Two Years before the Mast* (1840).[5]

As more people began to experience the ocean for recreation rather than for work, the fascination with sailor suits, shell collecting and seascape art was joined by an embrace of superstition, fantasy and lore that distracted the public from the actual ocean. The sea transformed into a place of mystery where spectre ships crossed paths with real ships and where mermaids and pirates lost their terrifying qualities. The real-life story of the brig *Mary Celeste* echoed the myth of the famous ghost ship, *Flying Dutchman*. Found adrift in the Atlantic in 1872 with its crew – including the captain's wife and baby – vanished, the lifeboat gone but no navigational equipment taken, and no indication from the log or the state of the vessel of any trouble, the mystery of the *Mary Celeste* not only fascinated journalists and newspaper readers but inspired artists and writers. Most famously, Arthur Conan Doyle published a fictional retelling, 'J. Habakuk Jephson's Statement' (1884), helping the tale remain alive in popular culture into the twentieth century.

Popular literature also featured in the rehabilitation of pirates from horrifying historical characters into heroic, or sometimes comic, figures. As the threat of piracy all but disappeared in the nineteenth century, novelists, playwrights and musicians created the dashing and colourful images of pirates that delighted and entertained ordinary people, both

Robert Louis Stevenson's map from his novel *Treasure Island* (1883), which created a popular image of pirates that has endured to the present.

landlubbers and sailors, children and adults. Robert Louis Stevenson, in his famous novel *Treasure Island* (1883), linked pirates with black schooners, treasure maps, islands and one-legged seamen with parrots perched on their shoulders. He single-handedly created the image of pirates that continues to animate popular culture today. *Treasure Island* inspired the well-known song 'Yo Ho Ho and a Bottle of Rum', which was a late nineteenth-century invention written after the novel, not a traditional sea song. Pirates became a popular and enduring subject of fiction for children beginning with J. M. Barrie's 1904 play *Peter Pan*, which introduced Captain Hook, so named for the iron hook that replaced his hand. Just as pirates were transformed, mermaids similarly morphed from a portent of doom to any sailor misfortunate enough to glimpse them into entertainment, decoration on yachting trophies or images in advertisements.

As landlubbers edged out sailors in generating cultural images of the ocean and sea-going, many gazed towards the sea in order to look inside themselves. Henry David Thoreau called the ocean 'a wilderness reaching around the globe', more like a jungle than civilized, industrialized land.[6] Once wilderness had been a fearful place, to be avoided or attacked and tamed. New to the nineteenth century was an appreciation of wilderness, which began with the Romantic embrace of its sublimely mysterious, dark and frightening elements and shifted to the desire to experience the intensity of personal encounters with the sea. Thoreau discovered the ocean from the sandy beaches of Cape Cod, witnessing the deadly wreck of the immigrant ship *St John* in 1849 yet nonetheless embracing ocean wilderness as a desirable antidote to civilization.

Appreciation for the sea metamorphosed into novel expressions of love for the sea. The American poet Emily Dickinson, a land-bound recluse, asserted her right to define human interaction with the sea. In an 1860 poem she invoked the feelings of 'intoxication' felt by 'an inland soul' who sets sail, claiming an experience of the sea that was unattainable to experienced sailors:

> Exultation is the going
> Of an inland soul to sea;
> Past the houses – past the headlands –
> Into deep Eternity –
> Bred as we, among the mountains,
> Can the sailor understand
> The divine intoxication
> Of first leagues out from land?[7]

At this time, when respectable men went to sea and inland souls dreamed with dry feet, the ocean took on a maternal character, and ships were imagined as female. Partly this reflected the fact that masculinity associated with seafaring became more pronounced when the industrial ideology of separate spheres for men and women strongly segregated social worlds ashore. Gendering of the sea included the metaphorical understanding of the sea as a womb of life.

The ocean's burgeoning role as a playground and a site for spiritual renewal accentuated its mental and imaginative contributions relative to its tangible – and continuing – economic and political ones. Literary scholars note the entrance of ocean metaphors and images, which came to dominate Western culture. The ebb and flow of tides described stages of life, while shipwrecks gained connotations from the political to the personal. Voyages emerged strongly as a metaphor for life's journey in art and literature, a trope as old as Homer's *Odyssey* and probably older but increasingly deployed by writers and artists with little or no experience of the sea. Consider, for example, the American Romantic painter Thomas Cole's series of canvases titled *Voyage of Life* (1842), which included childhood, youth, manhood and old age.

The transition from sail to steam accompanied the shift from oceanic work to recreation and the new-found participation of inlanders in the creation and consumption of maritime popular culture. All this set off a wave of nostalgia that historian John Gillis claims as part of a 'second discovery of the sea', similar in cultural influence to its first discovery, when explorers found sea routes between all the known lands of the globe in the fifteenth and sixteenth centuries.[8] Shanties were embraced as an appealing part of maritime culture, their popularity distracting from their function of easing the brutal work of hauling up anchors, nets or heavy sails with fewer hands. Some romantics gripped with nostalgia signed on to experience the last days of square-riggers, which lingered until the Second World War on the longest ocean routes in which steam could not yet compete, including the nitrate fertilizer and grain trades around Cape Horn and to Australia. Eric Newby based

The Last Grain Race (1956) on his 1938 voyage aboard the four-masted barque *Moshulu*. Young men similarly also sought out experiences on the last wooden whaleships. The naturalist Robert Cushman Murphy shipped on the whaling brig *Daisy* in 1912–13, later publishing *Logbook for Grace* (1947). The success of the German Flying P Line, with its fast and capacious four-masted barques whose names all began with the letter P, ensured that a number of square-riggers were available for preservation when revivalists began to preserve the memory and artefacts from the age of sail. The *Passat, Pamir, Peking* and *Padua* all survived the war and all but one function still, at least at dockside.

Older vessels, too, found keepers in the first half of the twentieth century dedicated to preserving maritime heritage, such as the 1765 naval vessel HMS *Victory*, the 1869 clipper *Cutty Sark*, the 1841 whaling barque *Charles W. Morgan*, and many others. The fishing schooner *Bluenose*, built in 1921, was equally designed for racing but became most important as an icon for Canadian maritime greatness. Maritime museums formed to house collections amassed by enthusiasts determined to save examples of the technology and culture of wooden ships and associated maritime work before it disappeared completely. A tradition of sail training emerged from traditional working vessels, leading to new missions for replica and modern 'tall ships'. That appellation came from the 1902 poem 'Sea Fever' by John Masefield, who sailed on naval and merchant vessels in the 1890s and coined the term that well serves a culture that cannot, and need no longer, distinguish between ships, brigs, schooners and other rigs. The husband-and-wife team Irving and Exy Johnson developed the model of character-building sail training starting in the 1930s by taking young men and women on long voyages aboard their series of vessels named *Yankee*.

The nostalgia for a grand maritime past resonated with readers who appreciated the yearning, romantic thread in Masefield's poem: 'I must go down to the seas again, for the call of the running tide / Is a wild call and a clear call that may not be denied'.[9] Maritime revivalism may also illuminate the rediscovery in the 1920s of Herman Melville's

Moby-Dick, unappreciated by critics and general readers when it appeared in 1851. Critics ascribe the novel's rediscovery to modernism, but it certainly appealed to readers who appreciated its homage to an evocative maritime past.

Western culture had long viewed the sea as a space outside society but the new wilderness role for the ocean emphasized the ocean as also outside of time. Thoreau expressed this characteristic as a contrast with the land:

> We do not associate the idea of antiquity with the ocean, nor wonder how it looked a thousand years ago, as we do of the land, for it was equally wild and unfathomable always . . . The aspect of the shore only has changed.[10]

Irving Johnson at the wheel of the schooner *Yankee* with his son and his wife Exy Johnson, returning to Gloucester in 1938 from a circumnavigation that taught young people how to sail during the voyage.

This sense of timelessness seemed to render the ocean impervious to human actions. George Gordon, Lord Byron, put it this way in his narrative poem *Childe Harold's Pilgrimage* (1818):

> Man marks the Earth with ruin – his control
> Stops with the shore; – upon the watery plain
> The wrecks are all thy deed, nor doth remain
> A shadow of man's ravage . . .[11]

Byron acknowledged the effect of human activities on land but considered the sea an untouchable realm.

The separation of the ocean from time reinforced ways of thinking about the sea and its resources set in motion by embrace of freedom of the seas. Imperial ideology and practice resulted in the conviction that the ocean and its resources should be exploited by those with the knowledge and power to do so. A timeless ocean that appeared impervious to people's actions resonated with perceptions of the limitless fish resources posited by Huxley. He and his contemporaries also viewed the depths as unchanging, full of ancient living species left from earlier geological eras. As a space outside society, the ocean was available to be defined by its utility for humans. The shift away from viewing the ocean as a place of work, and of economic and geopolitical importance, to the new cultural perception of the ocean as timeless and untouchable masked the nature and the scale of industrialization of the ocean.

THE SCALE OF FAMILIAR uses of the ocean – for warfare, transportation and fishing – escalated enormously during the twentieth century, and the two world wars significantly accentuated the tempo. As most wars had since the discovery of the seas, both involved significant fighting at sea and efforts to control ocean space. Submarines, used lethally in the First World War and in drastically expanded numbers in the Second, extended warfare from the surface into the ocean's third dimension.

Germany's adoption of unrestricted warfare during the First World War, including attacking vessels of neutral powers, inflicted enormous damage to shipping, transforming the depths into a source of hidden danger and reinforcing one of the benefits of freedom of the seas. This strategy contributed to the entry in 1917 of the until-then neutral U.S. into the First World War, although the U.S. later deployed it against Japanese shipping in the Second World War. The outcome of the latter conflict rested not only on the use of submarines but of aircraft carriers, destroyers, the many vessels that carried cargo and escorted convoys, and landing craft. The operation of all of these stimulated a massive effort to learn more about the ocean environment, including in the atmosphere, on the surface, at the boundaries of land and sea, and in the depths.

The massive new wartime yield of knowledge about the sea from the Second World War became available afterwards to transform traditional peacetime uses of the ocean. Technology for transporting materiel during wartime, for example, transformed the shipping industry. American entrepreneur Malcolm McLean developed an old idea for carrying freight in containers. Containers permitted armies to deal with the difficulties of different railroad gauges and the challenges of moving cargo quickly between land and sea. McLean, recognizing the potential for this idea, sold his family's large trucking company to buy a shipping company. In 1956 he converted two wartime tankers to carry steel containers he designed so that trucks could deliver and pick them up directly to and from ships.

McLean's new company realized enormous savings on the cost of loading and unloading cargo. These savings came at the expense of stevedores, large numbers of whom had packed cargo piece by piece or pallet by pallet into ships' holds. They lost their jobs. McLean's technological innovation was a standardized steel shipping box that could be loaded onto ships and secured for long voyages. He made his patented design freely available, a canny strategy because standardization encouraged dramatic growth of the intermodal system.

Malcolm McLean, inventor of the container shipping system, stands at a railing overlooking Port Newark in 1957.

An expanding global economy after the war fuelled the development of other specialized vessel designs from wartime shipbuilding techniques, including different designs for vehicles and for liquid or gas cargoes, as well as bulk vessels with booms to handle their own cargo and ro-ro's (for roll-on roll-off), which allowed wheeled cargo to be loaded and unloaded easily. This array of dedicated carriers enabled sea transport of four hundred times more cargo in the early twenty-first century compared to 1840. Decreased fuel consumption of 97 per cent since 1855 in conjunction with lowered labour costs, efficiencies of scale and market competition have kept transport costs low; for key commodities such as coal and oil, these hardly rose in the second half of the twentieth century. Today the market for raw materials and manufactured products is global. The shipping industry handles energy commodities; agricultural products; bulk commodities such as forest products, chemicals and minerals; and manufactured goods.

Ocean shipping, despite developed nations' utter dependence on it, has faded from view. Ninety per cent of the world's overseas freight is now moved by less than half a per cent of the population. Because containerization required space for the staging and storage of the boxes, new ports developed outside the traditional ones in major urban centres. Port Elizabeth, New Jersey, replaced Manhattan, and Oakland drew ship traffic away from San Francisco, while cargo vessels bypassed London in favour of the ports of Felixstowe in Britain and Rotterdam in the Netherlands. Such shifts left empty districts in port cities, eliminated jobs at docks and strongly promoted globalization of the world economy.

The ocean shipping industry not only underwent dramatic social, economic and technological change. It reshaped connections between people and oceans, and even altered coastal ecosystems. Reflecting the erasure from everyday consciousness of the enormous shipping industry, the shift from passenger liner to aircraft for trans-ocean travel similarly elided the ocean itself from the experience of travellers except as time and space to be traversed. Shipping came under far-reaching regulation intended to improve safety at sea and to address pollution issues. The increasing web of international agreements and national legislation governing global shipping drew the maritime realm more firmly under the same kinds of legal and political control long exerted over terrestrial industries. On vessels from wooden ships to gigantic tankers, countless unnoticed marine creatures had always travelled along with cargo, affixed to ships' bottoms or living in water taken on as ballast. San Francisco Bay, now the home of more invasive species than any other aquatic region, began receiving exotic species in large numbers during the Gold Rush of 1849, when vessels arrived from all over the world, many of which were abandoned in the harbour. Marine invasive species – blue migrants of a sort – transplanted by global shipping often displace native species and disrupt the ecosystems of their new homes.

The other traditional peacetime oceanic activity, fishing, likewise bore the imprint of changes inaugurated by the Second World War. During the hostilities, many fishing vessels had been commandeered, and fishing in areas such as the North Sea and North Atlantic abandoned or severely curtailed. Other fisheries continued despite the danger because of the essential food and materials provided for the war effort. Whaling continued in the eastern Pacific during the war, although the fleet left Antarctic waters after 1940. While concerns about overfishing and the perceived need for conservation carried over from before the war, hunger and the necessity to reconstruct Europe and Japan over-shadowed efforts to control the resumption of fisheries and instead promoted a derby approach to fishing.

Scientists and conservationists involved with the International Whaling Commission (IWC), formed in 1946, renewed pre-war inter-national efforts to protect whale populations for future use. The IWC aimed to balance science and industry, intending to use science to de-velop whaling and expand stocks. Instead, like so many fisheries, whaling companies faced evidence in the 1950s of declining stocks, a problem made worse by decreasing prices for oil as, among other factors, vege-table oil began to replace that from whales. Restriction of catches by lowering quotas proved unworkable due to features built into the IWC, compounded by the sense that there was insufficient knowledge upon which to base such measures. By 1960 whaling companies could not catch enough whales to fill reluctantly reduced quotas. By the middle of that decade, blue and humpback whale stocks were so low that hunt-ing for these species was closed. In the case of whaling, the dream of scientific management to maximize yield had failed.

Food shortages during and after the war prioritized state investment in fishing vessels, initially through the repurposing of vessels no longer needed for the war effort. Many such vessels entered traditional fisher-ies, but governments also funded experiments to expand the scale of harvesting and processing catches. The U.S. government, for example, supported the 1946 conversion of a 129-m (423-ft) military freighter into the *Pacific Explorer*, intended to act as a mother ship to smaller purse seiners, which could either freeze the catch and transport it to shore-side canneries, or can fish while at sea. For a variety of reasons, this trial proved less successful than a British whaling firm's attempt to transform a 67-m (220-ft) minesweeper into the fishing vessel *Fairfree*, outfitted with an experimental freezing system and sent out on test voyages in 1947. By the mid-1950s larger British and Soviet purpose-built factory trawlers began fishing and processing enormous catches, and were able to remain at sea for months at a time. Soon many developed nations representing both sides of the Iron Curtain built and operated factory processing ships, exploiting virtually all existing high seas fisheries. Industrial fisheries trawled for bottom fish such as cod; purse-seined for pelagic schooling fish such as herring; long-lined for highly migratory, high seas species such as tuna; and employed huge new mid-water trawls. Together the new suites of fishing technology enabled access to virtually all of the horizontal and vertical extent of the ocean's continental shelves and slopes.

Dramatic post-war catch increases resulted from the massively increased fishing effort, reinforcing the conviction that the sea's resources were essentially without limit. By contrast, pre-industrial fishers had recognized declines of fish catches in response to intensive fishing and, at times, attempted to limit fishing. Post-war optimism rooted in the rapidly expanding field of fisheries science fuelled ambition to expand fisheries, not limit them. Restrictive measures, such as rules about the size of mesh in fishing nets or the length of fish landed, were discussed seriously in Europe before the First World War but only implemented later and with great reluctance. In most parts of the world, restriction of ocean fisheries came only after major fish stock collapses, via attempts to impose national quotas and regulation of fishing effort.

Institutions with the power to regulate Atlantic fisheries emerged after the Second World War. European fisheries scientists created the wartime 'Second Great Fishing Experiment', an opportunity to institute regulations before fishing resumed, with the goal of sustaining higher levels of stocks that had recovered during the hostilities. Before the war, the English scientist Michael Graham demonstrated his 'Great Law of Fishing' – namely, that unlimited fishing becomes unprofitable, arguing that the industry would benefit from restriction as much as the fish would.[12] Although post-war food shortages made governments reluctant to limit fishing, northern European nations did convene the 1946 London Overfishing Conference and created the Permanent Commission with the power to enact international fisheries regulations. (In 1963, a decade after its formation, it became the Northeast Atlantic Fisheries Commission.) This institution established an advisory role for science in fisheries management, and was soon joined by other similar international fisheries management bodies.

From wartime science and technology, fisheries scientists gained a tool to help them advise managers. Graham's military service with artillery ballistics introduced him to the mathematics of targeting, which after his peacetime return to fisheries science prompted him to hire a biologist and a mathematician to work together to quantify the dynamics of stocks in response to fishing. In 1957, Raymond Beverton and Sidney Holt published what a generation of their colleagues referred to as 'the Bible', *On the Dynamics of Exploited Fish Populations*, which gave scientists a simple model that could be used to estimate the yield of a fishery under various conditions. This tool provided valuable information for those tasked with managing, and perhaps regulating, fisheries. Embraced by the Permanent Commission charged with oversight of the heavily utilized Atlantic fisheries, this model infused scientists and managers with confidence that human control over the ocean through prediction of catches was within grasp. Crashes of British herring fisheries in the late 1950s motivated scientists to redouble efforts to apply their vastly increased knowledge of the oceans to help the fishing industry find new populations and species to exploit.

The Pacific presented a different outlook. After the war, the u.s. and Japan in particular understood fisheries as intimately linked to geopolitical ambitions. Two fisheries in particular, for tuna and for salmon, dominated international relations. The San Diego tuna fleet moved south to find fish to feed canneries when u.s. stocks faltered, provoking the ire of first Mexico and then Central and South American nations who resented Americans fishing off their coasts. The u.s., though, wanted to find some means to prevent Japanese fishers from taking salmon from Bristol Bay, the easternmost part of the Bering Sea off Alaska. An energetic and entrepreneurial fisheries scientist, Wilbert Chapman, created a fisheries position in the U.S. State Department and used that role to develop a strategy to achieve the diametrically opposed goals of fishing for tuna far off American shores while preventing the Japanese from fishing for salmon close to the U.S.

In contrast to Michael Graham, Chapman believed that less fishing was wasteful, but he shared with Graham the conviction that science offered the key to expanding yields. He seized on the concept of Maximum Sustainable Yield (MSY), a level of fishing to be determined by science that would ensure the greatest possible use of the resource. He made this concept part of American fisheries policy in 1949. Channelling Huxley, Chapman argued that fisheries must remain open unless scientific research could demonstrate that a coastal nation needed to close them for conservation purposes, neatly laying the foundation for Americans to fish tuna off foreign shores and reserve Bristol Bay salmon for themselves. Although MSY sounded scientific, Chapman did not publish about it in a scientific journal or discuss the concept with reference to scientific literature. MSY was a political concept first, and only later did fisheries scientists create mathematical formulas to establish levels of maximum fishing. MSY was based on several assumptions, including that fishing is good for stocks because removing large fish leaves food for smaller ones; that stocks are resilient even under constant fishing pressure; and that free markets will act to protect stocks.

Global catches skyrocketed in the twentieth century. At the start of the First World War, worldwide landings were 9 million tons, jumping to 20.7 by the eve of the Second World War. Catches rose throughout the 1950s and 1960s, hitting 27.4 million tonnes in 1961 and spiking to about 55 in 1970. Post-war innovations included the application of new materials to traditional, even ancient, fishing methods, such as the use of nylon instead of natural fibres for nets. The wartime technology of sonar, developed for submarines, made the traditional purse seine method of capturing pelagic fish such as herring unerringly deadly. At times used in conjunction with spotter airplanes, which scouted for schools of fish, echo-sounders allowed fishers to encircle an entire school, and power blocks ensured that huge nets could be recovered so that no fish got away. The breathtaking expansion of catches facilitated by factory-style fishing embodied a change in scale, to be sure, but it also made every part of the sea reachable by technologies capable of harvesting with lethal efficiency. Eventually, with the inevitable decline of catches from one area, these vessels sought and exploited new stocks of fish further away from their home ports and in deeper waters, increasingly often in the 1960s and 1970s fishing heavily off the shores of poor, newly independent countries in Asia and Africa.

Despite Graham's conviction that certain fisheries demanded regulation, even he judged the ocean the source of untouched wealth and a place unchangeable by human activities. 'It would seem', he wrote in 1956, 'that here at the beginning and the end is the great matrix that man can hardly sully and cannot despoil.'[13] Industrialized fisheries managed by scientific experts spread around the globe as developed nations took to heart the adage about teaching a person to fish: give someone a fish and you feed her for a day; teach her to fish and she can feed herself for a lifetime. Reflecting Cold War political tensions and mounting concerns about global overpopulation, technical assistance to developing nations targeted not only agriculture but fisheries. Modern trawlers and gear were introduced in areas of the world where subsistence fishing using ancient methods was the norm. Fisheries scientists taught local technocrats in developing nations how to use the new models to manage expanding industrial fisheries, and such science education was intended to foster the intertwined goals of modernizing fisheries, economies and societies. Although technology and science supported the dramatic expansion of fishing and other traditional uses of the ocean in the twentieth century, the relationship between people and oceans remained a human one embedded in politics, ideology and ambition.

IN 1968 GARRETT HARDIN articulated 'the tragedy of the commons' to explain the tendency towards overuse of a common resource, such as a pasture, due to individual self-interest.[14] Referring briefly to the freedom of the seas, he asserted that fish and whales suffered from treatment as a common resource, but historian Carmel Finley aptly points out that governments and scientists put in place policies that shaped the global fishing and whaling industries into an enterprise that was anything but governed by the behaviour of individuals.[15] Tragically indeed, Hardin's thesis implied the unlikelihood of controlling the commons, and its most typical interpretations masked the role of policies and beliefs that promoted overuse of ocean resources.

Other traditional uses of the ocean skyrocketed alongside fishing, including the extension of warfare into the depths and the continuation of submarine strategies in the Cold War, as well as the dramatic expansion of global shipping. All of these activities tightened the connections between people and the ocean. Ironically, the rising trajectory of ocean exploitation was matched by declining awareness of and involvement in work associated with the sea. As the ocean transformed into a playground and retreat in the nineteenth and twentieth centuries, nostalgia for a lost maritime past fixed the ocean in a perpetually pre-industrial moment. Mariners, cartographers and readers in the several centuries before had understood the ocean as a site of human activity, though often outside of society. Once change was rendered inconceivable, the ocean stood outside history as well as society. An optimistic future was envisioned for the timeless sea as traditional maritime activities were joined by a suite of new uses for the ocean that reflected post-war confidence in the possibility of knowing and controlling the ocean.

SIX

Ocean Frontier

Obviously man has to enter the sea. There is no choice in the matter.
The human population is increasing so rapidly and land resources
are being depleted as such a rate, that we must take sustenance
from the great cornucopia.

– Jacques Cousteau, *The Silent World* (1953)

Picture this: a flat, broad expanse of open space that holds unlimited potential for agriculture. Wide, fertile pastures await harvest
and promise sustenance for vast herds that might provide meat or milk
for growing human populations. The forests at the margins of these
great plains harbour animals whose glossy pelts have long allowed capital
to flow outwards from settled and industrialized parts of the world.
Underground lies uncountable mineral wealth available for the taking
as soon as engineers can devise new technologies to reach it. The vast
third dimension of this region is the domain of unnumbered birds in
the air above and an unlimited supply of fish in its depths.

This description, until the fish that is, might have evoked the American
West at the so-called 'opening of the frontier'. Instead it expresses the
view of entrepreneurs and investors of the mid-twentieth century who
eagerly embraced the potential of the open sea for extractive industries,
cash crops and livestock, just as their nineteenth-century counterparts
had for western lands. While the ocean had long served as an arena for
war and a font of resources, the post-Second World War ocean came
to be viewed through the cultural prism of the 'frontier', particularly in
the minds of the American scientists and engineers who participated
in the meteoric growth of oceanography, marine sciences and ocean
engineering during and after the war. The metaphor of 'frontier' equally
appealed to those who funded ocean science, to entrepreneurs who
hoped to create a high-tech ocean industry sector parallel to the lucrative

aerospace industry and to writers, readers and recreational users of the ocean who eagerly embraced the idea of the sea as a 'frontier' they could explore personally.

THE WAY THAT THE TERM 'frontier' was used to describe the ocean drew inspiration from the famous – one might even say infamous – associations with that term forged by the U.S. historian Frederick Jackson Turner at the end of the nineteenth century. Many characteristics and outcomes Turner associated with the American western frontier were embraced to express the perceived bounty of the sea by people perhaps best called 'ocean boosters'. To offer a compressed version of Turner's argument, the first European settlers to move into the American West were trappers and traders, followed by cattlemen and miners. Next came subsistence farmers. Eventually farming grew more intensive and settlements larger, until cities and industries formed. The inevitable outcome, as described by Turner (and predicted for the ocean by post-war boosters), included access to almost-endless food resources, fantastic wealth from extractive and productive industries, new living space and the continued development of individuals and of political and social institutions. Although historians had discredited much about the content and implications of Turner's frontier thesis by the mid-twentieth century, his characterization of the American West proved irresistible to ocean boosters. Turner's West produced wealth, provided resources and territory for expansion, fostered innovation and technological development, and even promoted individualism, self-improvement, democracy and progress.

Ocean boosters believed the ocean similarly would provide the material resources and environmental challenges that would prompt continued progress into the future. One such commentator, the engineer and popular writer Seabrook Hull, enumerated in 1964 the ways the ocean could be considered a frontier:

Of the two great frontiers, space and the ocean, being opened up in the 20th Century, only the ocean is close, tangible, and of direct personal significance to every man, woman, and child on the face of the globe. Another war might be won or lost in its depths, rather than in outer space. It is a cornucopia of raw materials for man's industries, food for his stomach, health for his body, challenges to his mind, and inspiration to his soul.[1]

What kind of frontier was the ocean? Hull spoke for his generation, who believed that the sea promised food and other material resources; challenges to inspire novel and productive industries; and fulfilment of intellectual and spiritual needs as well.

References to the ocean as frontier appeared soon after the war's end, most frequently alluding to either economic potential or new scientific knowledge. In November 1953 the American Association for the Advancement of Science held a special session on 'The Sea Frontier', proposed by James B. Conant, long-time Harvard University president who served as an adviser to the National Science Foundation and the Atomic Energy Commission. Co-organized by an oceanographer from Woods Hole Oceanographic Institution and an engineer from the Massachusetts Institute of Technology, its topics ranged from geology of the ocean basins, to the productivity and biological resources of the sea, and to the potential for extracting resources such as fresh water or minerals. On the popular front, a 1954 advertisement in *Life* magazine placed by the American Petroleum Institute added the element of struggle against nature: 'In the open waters of the Gulf of Mexico, against every hazard of wind, wave, and sudden storm, sea-going oilmen are opening up a new American frontier.'[2]

Economic development was by no means the only reason for calling the ocean a frontier, although it did lend heft to more symbolic uses of the metaphor as applied to exploration and scientific discovery. Marine geologists and geophysicists adopted the language of frontier to express the idea that the sea floor, especially in the deep sea, was 'the last geographic frontier' because of the gigantic peaks and enormous mountain ranges, the deep basins and trenches, and the wide fault zones being revealed by geophysical research, with 'still others waiting to be discovered.'[3] 'Get Wet Young Man!' served as the rallying cry for the author of the 1968 young adult book *Explorers of the Deep*, channelling the slogan ascribed to the nineteenth-century newspaperman Horace Greeley, 'Go West Young Man!'[4] Enthusiasm for the ocean's potential ran high. Boosters believed that its resources were, for all practical purposes, limitless. This thought reassured a world that believed it must soon look beyond Earth's landmasses for the means to sustain civilization. Many futurists, of course, looked to outer space, but the ocean was also embraced as the 'Last Frontier', the title or subtitle of numerous popular books published in the 1960s.

Exploiting the ocean's vast resources depended on anticipated advances in science and engineering. Like space, the depths were a technological frontier. Innovations promised to enable economical use

City under the Sea, painted by German graphic artist and futurist Klaus Bürgle in 1964.

of food and mineral resources from the sea, likely involving human divers living and working from underwater bases. Scientists and futurists forged the emerging vision of the ocean as frontier, taking their cue at least in part from the trailblazing 1945 report to President Franklin D. Roosevelt, 'Science: The Endless Frontier', which laid the groundwork for civilian funding of American science through the National Science Foundation.[5] This report portrayed science as closely resembling the Turnerian western frontier, predicting that investment in science would yield jobs, health and prosperity, as well as innovation, democracy and progress. Although historians by then dismissed Turner's thesis as applied to the history of the American West, the arc of his argument was transferred to science, and to the ocean's depths.

The strong association of science with the ocean during and after the war fuelled the adoption of the frontier label. Not only did oceanography grow enormously, but the technology to comprehend and control the ocean appeared to be on the horizon. Existing histories of ocean science describe the formation of fields or subfields that are well articulated today, such as physical oceanography or fisheries biology, but from the perspective of the 1960s, science of the sea seemed poised to combine physics and biology with engineering, human physiology and archaeology to support a new human relationship with the sea, particularly its depths. Plans for research facilities reveal dreams of the integration of science, industry, aquaculture, government and recreation. The new blueprints for ocean science addressed every possible aspect of how people might work, play and live on and under the sea.

The optimism and enthusiasm for the ocean as a new frontier surged from the pages of a pair of books published in 1960: *The Frontiers of the Sea*, by Robert C. Cowen, and *The Challenge of the Sea*, by Arthur C. Clarke. Clarke was better known as a science fiction writer and for his successful prediction of communications satellites, but he also published futuristic non-fiction and, after learning to scuba dive in the early 1950s, spent about a decade of his life obsessed with diving, diving-related businesses and writing about the ocean, often comparing its exploration

Arthur C. Clarke diving, probably in Ceylon (now Sri Lanka), *c.* 1955, photographed by his diving and business partner on numerous underwater ventures, Mike Wilson.

to that of space. Cowen earned a Master's degree in meteorology from MIT in the early 1950s and then left academia to help the *Christian Science Monitor* improve its science coverage. Near the beginning of his long career, which was studded with science writing prizes, he wrote his ocean book, which surveyed the wealth expected to be drawn from the sea, ranging from oil to minerals, and to protein from fish and plankton. Energy, fresh water and other more remote possibilities were also featured. The year 1960 saw the bathyscaphe *Trieste* carry two men to the deepest part of the ocean, the Challenger Deep in the North Pacific Ocean's Mariana Trench. That technological accomplishment seemed to confirm that humans were on the verge of conquering the depths.

Clarke and Cowan rode the crest of a wave of similar books extolling the wealth that would soon be extracted from the sea or predicting the imminent ability of the ocean's resources to improve life across the globe. Titles such as *New Worlds of Oceanography* (1965) suggested the sea's parallel to North and South America as new continents discovered by hemmed-in Renaissance Europeans. The word 'frontier' appeared frequently, as in *Undersea Frontiers* (1968), *Oceanography: The Last Frontier* (1973), *Oceans: Our Continuing Frontier* (1976) and numerous others. Titles such as *The Bountiful Sea* (1964) and *The Riches of the Sea* (1967) evoked the many sources of wealth expected from the ocean, while others focused on particular assets, as *The Mineral Resources of the Sea* (1965) and *Farming the Sea* (1969) did. An almost equal number of volumes for young adult readers appeared, with similar titles and subtitles: *The Challenge of the Deep Frontier* (1967), *Turn to the Sea* (1962), *Underwater World, Explorers of the Deep* (1968) and *Hydrospace: Frontier beneath the Sea* (1966).

To authors of these books, the sea appeared poised to metamorphose into territory available for human expansion, promising to feed a growing population and to provide minerals, fresh water and energy, or perhaps even, some day, living space. 'Hydrospace' would not only serve as a zone for resource extraction but as a site for new industries providing novel kinds of jobs for future generations.

MANY ANTICIPATED OCEAN enterprises involved the ocean's third dimension and the efforts of 'aquanauts' working in the sea. Commercial availability of the self-contained breathing apparatus transformed what emerged from the Second World War as specialized military hardware for daring and heroic frogmen into a tool permitting ordinary people to go underwater. New materials, scientific understanding of the ocean and engineering and management knowledge fuelled fantastic visions for exploiting the sea. Scientists, engineers and entrepreneurs expected the anticipated ocean technology industrial sector to rival the aerospace industry.

Even the more traditional uses of the ocean swung into the sights of boosters. In 1959, a hovercraft successfully crossed the English Channel, inspiring great hopes for this novel technology. Such craft would free ship owners and captains from concern about the reefs and shoals that endangered traditional vessels, opening up many areas for navigation that prudent mariners must avoid and some places that are not at all accessible to ships, including ice and snow fields, farmland, swamps and even molten lava.

Like containerization, hovercraft technology took aim at the problem of moving cargo across the land/sea interface by simply carrying on from the water across dry land, stopping only at the cargo's destination. Hovercraft cargo transport would render unnecessary the infrastructure of great highway systems, at that time under construction in the U.S. It would also extend to the land a similar concept of freedom of the seas – that is, if significant political obstacles could be addressed and as long as freedom of the seas remained the customary legal regime for the world's ocean. Neither happened and the hovercraft never fulfilled the promise its boosters imagined.

Beneath the hovercraft, boosters also imagined a potential cargo-carrying transportation system operating in the ocean's vast third dimension, free to move anywhere within this realm that had long been considered the territory of no single nation. Submarines had played a critical role in two world wars, emerging as a foundational element for the military operations and strategy of the superpowers and a few other developed nations. Nuclear submarines, starting with the acclaimed USS *Nautilus*, launched in 1954, made submerged operations possible for months at a time. Experimental submersibles of various kinds broke records for the depths to which underwater vehicles could function. Against this backdrop, plans for submarine cargo transportation emerged as an adjunct to the nascent offshore oil drilling industry.

Transporting liquids or gases under the sea appeared to offer several advantages. First, submarines could travel along direct routes, avoiding surface phenomena such as storms, adverse currents or waves, or maritime traffic. Second, the pressure at great depths was an asset for resources such as gases that must be stored or transported under pressure. Similarly, the undersea environment kept such resources free from temperature fluctuations and provided protection from oxidation for resources that deteriorated with exposure to air. A significant logistical benefit derived from the fact that tankers often returned from oil deliveries empty. If oil or other liquids were instead transported undersea in giant rubber bags towed by a submarine engine, then, at the destination, the bags could simply be folded and stored in a small space, saving fuel on the return trip.

Cargo transport was not the only traditional maritime undertaking that ocean boosters imagined as a futuristic, undersea activity. Food from the sea had always been important, but plans were laid to shift away from haphazard, casual exploitation of the ocean. Chasing fish, even using sonar technology newly adapted from wartime defence research, reflected what was understood as an outmoded hunting model in a world that boosters pointed out had long since, on land, turned to farming. Target organisms for aquaculture included seaweed, shellfish, shrimp, lobsters and finfish of all types that inhabited the sea bottom, the surface waters and the depths in between. A self-described scrupulous journalist intent on presenting science rather than science fiction penned a 1969 book on farming the sea that discussed only actual experiments, such as his own efforts to farm crabs and lobsters, as well as attempts to build and study artificial reefs and to develop methods to farm shrimp. His 'logical conclusion' was that sea farmers must eventually live in the ocean.[6]

Discussions of farming the sea inevitably invoked the spectre of global population growth and the attendant challenge of feeding so many people. Scientists and technical experts often articulated the belief that converting plankton into some kind of protein source offered an appropriate solution for the hungry citizens of developing nations, although the same scientists tended to imagine their own country's citizens eating farmed salmon instead. Today's fish farming, involving a handful of species and utilizing areas near shore, represents a pale reflection of the expansive 1960s vision for ambitious projects involving fertilizing the open ocean, ranching fish on the high seas, massive-scale harvesting of plankton and even tending herds of whales.

The firm understanding of whales as a food resource inspired 1960s boosters to advocate the abandonment of hunting them, in favour of farming. Proposals for ranching whales, while seeming perhaps fanciful today, emerged from the logic that whales represented large sources for protein and other commodities, combined with the worry that the modern whaling industry would soon extirpate them. Like bison in the American West, to which they were frequently compared, the great whales might be saved for future generations by raising them in captivity. Gifford Pinchot Jr, the son of the famous architect of the American conservation movement Gifford Pinchot, proposed that coral atolls could provide natural fences, while futurists anticipated bubble nets to enclose whales inside enormous swaths of the ocean extending from the polar regions where they feed, to the tropics where they migrate to give birth. In his non-fiction work, Clarke confidently predicted that future generations would associate the word 'farm' with the Antarctic Ocean, whose whale populations promised 'the richest of harvests.'[7]

Clarke's 1957 novel, *The Deep Range*, though it seems nearly ridiculous today, was grounded in plausible, if optimistic, science. In it, a future Earth has solved the challenge of feeding the world's enormous population by farming the sea, which is divided into areas for plankton harvesting with gigantic floating reapers and those set aside for ranching whales. Modern cowboys working from small submarines tend and drive the herds, protecting them from voracious killer whales and keeping the bubble fences in good repair to prevent escaped whales from stealing plankton from the areas designated for its harvest. An idea emerges to train killer whales like sheepdogs, in order to herd the great whales, reflecting achievements in training porpoises.

The story begins, as any good maritime tale does, with the training of a neophyte whale warden in order to introduce the reader to this fantastic world. The new warden, Walter Franklin, had a failed career in space before being sent to work in Earth's ocean, a biographical detail that reappears at the denouement. Clarke's 'deep range' strongly resembles Turner's western frontier, including the preoccupation with its role in promoting human development. As many ocean boosters did, Clarke believed the sea to offer not only greater challenge than outer space but more immediately accessible resources as well.

In a confrontation with a Buddhist leader who speaks against the killing of the whales, Franklin, by then director of the Whale Bureau, surprisingly agrees to pursue the possibility of exploiting whales for milk instead of meat. This idea, like most of Clarke's ideas, resonated with contemporary scientific inquiry. Physiologists studied the composition of porpoise milk as early as 1940; by the 1960s they had discovered the milk of several species of seals to be 50% fat, prompting field researchers to visit Pacific breeding sites for elephant seals, walruses and various species of sea lions and seals in order to collect milk samples for analysis. The leap to imagine human food uses for whale milk was not any longer than the exercise of predicting plankton burgers. Both of these prospective foods from the sea introduced a chapter of the young adult book *The Sea: A New Frontier*, produced by educators in California working with the Scripps Institution of Oceanography in 1967.

WILDLY AMBITIOUS PROPOSALS for exploiting living marine resources, even with slender ties to on-the-shelf science or technology, were guaranteed a hearing with an audience who thought of the ocean's third dimension as a frontier. Ideas such as fertilizing vast parts of the ocean to increase productivity or installing nuclear reactors on the sea floor to create artificial upwelling zones were seriously proposed, the latter suggested for a pilot study by a committee of the National Academy of Sciences in 1959.

Far from viewing the ocean's resources as limited, post-war ocean boosters expressed confidence that the ocean represented 'a boundless, inexhaustible storehouse of the material stuff of civilization', in the words of the chief investigator for an economic analysis of ocean floor mining.[8] In 1961, the newly-elected u.s. president John F. Kennedy declared, 'Knowledge of the ocean is more than a matter of curiosity. Our very survival may hinge upon it.'[9] In this era of nuclear fear, the army-officer-turned-journalist Vernon Pizer wrote reassuringly in 1967 that if all terrestrial resources were suddenly to disappear, 'man could find in the world ocean virtually everything he needs to maintain himself in reasonable comfort.'[10]

'Everything' meant exactly that. While humans had relied for aeons on the sea for fish, salt, transportation and other resources ranging from shells and coral to rare amber and precious pearls, post-war experts pointed to new resources that might augment or replace land-based sources of important metals and minerals. Fresh water might flow from desalination plants to satisfy the needs of dry coastal areas. Energy from tides or thermal masses in the sea might be harnessed. Pharmacologists anticipated tapping the ocean as a new source of drugs. Oil, diamonds and sulphur seemed potential oceanic resources as well.

Chemists enthusiastically enumerated the contents of a bucket, or an acre, of seawater, finding at least 32 elements and noting that even trace amounts of an element such as copper or gold added up to significant potential profits, if only the means could be found for extracting it from seawater. Important industrial raw materials, including salt, potassium and bromine, were found in plentiful supply. Those materials readily and cheaply obtainable on land such as copper and potassium, boosters agreed, would reside unexploited in the sea for the near future. Soon however, several enterprises emerged to represent the leading edge of a brand-new set of industries able to extract wealth from the sea.

Bromine was the first element to be recovered in industrial quantities from seawater. Traditionally it came indirectly from the sea, from ash leftover from burning seaweed or deposits left by ancient seas, to be used in the manufacture of dyes and for photography and medicine. The Dow Chemical Company learned how to extract it from brine wells to increase production when demand spiked after the discovery that ethylene dibromide dissolved tetra-ethyl lead, creating an anti-knock agent for gasoline to protect internal combustion engines. Efforts to scale up this process to obtain the bromine from seawater resulted by 1934 in a large plant handling 227,000,000 l (60,000,000 gal.) each day that transformed bromine from a rare element to a moderately priced industrial input.

A second element obtained from seawater, magnesium, underwent a similar transformation in response to wartime demands. Magnesium was used in incendiary bombs, flares and various other military applications. Estimates of the amount needed for the Allied war effort were dramatically revised upwards after the discovery that Germans employed it in the metal used to build planes to make them lighter. Under great secrecy, Britain and the U.S. began developing production processes to extract magnesium from the sea. The 2,400 tonnes produced in the U.S. in 1938 grew to 240,000 tonnes in 1943, while the price tumbled from about $4 a pound in 1916 to about 20¢. So much magnesium was being produced by the war's end that some observers doubted whether civilian uses would expand to utilize the new capacity.

Companies that recognized the potential for exploitation of ocean resources invested in research and development. The 1966 President's Science Advisory Committee report on 'Effective Use of the Sea' enumerated for the U.S. a $10 billion investment in offshore oil and gas by the petroleum industry, a $900 million sand and gravel dredging industry, and sales of $45 million from sea-floor mining of sulphur and oyster shell. By the mid-1960s, 16 per cent of the petroleum used by the free world came from sea-floor wells. Predictions for the following decade anticipated that figure would double. Across the globe, more than a hundred companies operated in the waters of sixty countries while, in U.S. waters, $700 million worth of oil and gas was extracted annually.

Technology from the emerging offshore oil industry played a key role in one of the most breathtakingly ambitious oceanic projects of the post-war period, the attempt to drill through the Earth's crust at its thinnest point at the deep-sea floor with the goal of reaching the mantle. The boundary between crust and mantle, the Mohorovicic

Illustration of an imagined future undersea oil industry, in which drilling, refining, storage and transportation of oil was expected to take place entirely under the surface of the ocean, supported by developments in technology and saturation diving.

Discontinuity, or Moho, lent its name to the 'Mohole' project. Suggested in 1957 in part to shed light on the still-controversial question of continental drift, Mohole received funding from the National Science Foundation for experimental drilling in 1961 by a converted surplus naval freight barge, CUSS I, that a consortium of oil companies converted to develop offshore drilling capability. The operational phase of the project was cancelled due to escalating costs in 1966 without reaching the Moho boundary, but the project demonstrated the feasibility of deep-ocean drilling for geology, with applications also for the oil industry, and contributed dynamic positioning technology, which was widely adopted to keep ships in a single position for numerous industrial and research purposes. Oil companies experimented with drilling from rigs, surface ships and even semi-submersible platforms that could be towed to the site and anchored into place. Some of the work associated with laying pipelines and conducting drilling operations had to be done underwater. In the Gulf of Mexico, ambitions to drill in deeper and deeper water inspired the initially small Taylor Diving and Salvage Company to experiment with underwater construction and diving equipment and techniques. Drawing on expertise and experience from the u.s. Navy Experimental Diving Unit, Taylor Diving's leaders achieved the ability to operate in depths of 30 to 60 m (100 to 200 ft) between the mid- and late 1960s, pushing deeper in the following decade and setting international industry standards.

The company's success rested on technological innovations such as the first industrial recompression chamber and equipment for aligning and welding pipelines underwater, along with an experimental approach to building capacity for divers to work in greater depths for longer times. Under the pressure existing in deeper waters, gases in the lungs of a human diver dissolve into the blood and tissues, so that a diver who returns too quickly to the surface faces painful, or even deadly, decompression sickness, more commonly known as 'the bends'. The u.s. Navy developed decompression tables in the 1930s based on work by the British physiologist John Scott Haldane studying underground workers afflicted by the bends. Once tissues are saturated with nitrogen, the decompression time required for divers to return safely to the surface never increases, even if dive time is extended. Oil companies, wanting to increase the proportion of productive bottom time relative to decompression time, eagerly anticipated the development of saturation diving.

Soon engineers and ocean boosters envisioned entire oil-drilling operations under the sea, including mobile drilling rigs and refinery operations, giant undersea storage facilities, submerged pipelines, nuclear cargo submarines transporting rubber bags filled with oil and a working and living compound for the undersea workers. The Northrop Corporation invested in engineering studies for an undersea complex intended as an oil-drilling work camp to be installed at depths up to 300 m (1,000 ft). An elevator system would transport workers to the top section of a three-level structure. The middle level had five wings radiating outwards like the arms of a starfish, containing sleeping, eating and recreation areas as well as a laboratory. The facility would house fifty workers at once. The lower level held the electrical, air and other systems as well as the entrance to tunnels leading to three oil-drilling structures. The estimated price tag of $6.5 million equalled the cost of a research vessel at the time.

The ocean boosters who reported on the Northrop design and other oil industry undersea efforts believed that the U.S. Navy was involved in similar experiments. One claimed that the Navy had constructed a petroleum storage tank on the ocean floor in the Gulf of Mexico that had a capacity of 189,270 l (50,000 gallons). Possibly such a facility could support an underwater defence base of the sort that a few ocean journalists claimed the Navy intended to install in strategic locations around the globe.

LESS VEILED IN SECRECY was the U.S. Navy's saturation diving programme led by Dr George Bond, affectionately called 'Papa Topside' by the divers assigned to work with him. Bond's work in the Navy initially focused on improving the odds for escape from disabled submarines. His interest in the sea derived from his conviction that human survival would depend on developing the capability to operate underwater, to '[expand] man's ability to utilize the products of the marine biosphere which make up nearly three quarters of our Earth.'[11] He named the experimental programme to develop saturation diving 'Project Genesis', viewing his work as extending human dominion over the sea as promised in the Old Testament origin story. The Genesis 'dives' took place in laboratories on land in 1962 and 1963, testing human ability to breathe helium and to survive the pressures at depths of 30 and almost 60 m (100 and 200 ft), but they inspired a number of experimental saturation diving experiments in the sea.

The American inventor Edwin A. Link conducted the first opensea test of Genesis experiment results in Villefranche Bay in the French Riviera. Famous for creating the flight simulator that launched an entire industry, Link turned his energies to improving diving technology after he developed an interest in underwater archaeology. With support from the National Geographic Society and the Smithsonian Institution, he developed a diving system to enable people to work for long periods on the sea floor. He himself entered the narrow cylindrical pressurized chamber that was lowered to the sea floor for the first trials of two to eight hours of bottom time. Success emboldened Link to plan the 1962 'Man-in-the-Sea I' experiment, in which the Belgian underwater archaeologist and treasure hunter, Robert Sténuit, spent 24 hours in Link's cylinder at a depth of 60 m (200 ft), making a number of forays for simulated work, thus becoming the first aquanaut.

Mere days after Sténuit's achievement and only 160 km (100 mi.) away, off the coast of Marseilles, Jacques Cousteau began an experiment, 'Pre Continent I' or 'Conshelf I', to send people to live on the sea floor for one week. Cousteau had co-invented the self-contained underwater breathing apparatus in the 1940s, and marketed it energetically after the war. With French government funding, he anchored a cylindrical steel habitat to the sea floor at a depth of 10 m (33 ft) to provide housing and a base of operations for two divers who worked for five hours each day building fish pens, photographing fish and surveying underwater topography. Back in the habitat, named Diogenes, the men ate, slept and enjoyed a radio, phonograph, telephone and closed-circuit television. Judged a success, the programme led directly to 'Conshelf II' the following year in June.

Conshelf II was far more than merely a test of a habitat, although its primary goal was to see if five men could live and work on the sea floor for four weeks at the relatively modest depth of 10 m (33 ft) and working down to 18 m (60 ft). In addition, two divers would spend a week in a smaller outpost habitat 25 m (82 ft) deep and working in water almost as deep as 50 m (160 ft). A French petroleum consortium eagerly supported the programme, paying half the costs out of keen interest in the prospect that Cousteau's divers might demonstrate the feasibility of keeping divers on the sea floor for long periods of work. The aquanauts constructed a hangar underwater for Cousteau's diving saucer, which was used extensively in the research, filming and other activities carried out during forays from the habitat. A contrast to its spartan predecessor,

Starfish House featured separate dining, living, sleeping and working areas and was air-conditioned. A chef numbered among the crew, and one diver brought along his parrot, Claude. Starfish House, in fact, was as much movie set as work base, reflecting the other half of Cousteau's sponsorship, from Columbia Pictures for the documentary film made of the habitat's installation and use, *World Without Sun*.

Two months before the start of Conshelf II, the U.S. Navy submarine *Thresher* sank on 10 April 1963 in 2,600 m (8,400 ft) of water, lending urgency and a renewed military motive for solving the challenges of placing and keeping humans in deep water. Cousteau's projects had demonstrated successful saturation diving at depths reachable by surface divers. A pair of efforts in 1964 tested the concept at greater depths for longer time periods. Link put two aquanauts at 122 m (400 ft) off the Bahamas for 49 hours, during which time they used an inflatable habitat as a base and had access to a submersible decompression chamber. The U.S. Navy's Sealab I effort, overseen by George Bond, kept four divers in the 7-m (24-ft) habitat at 60 m (193 ft) for eleven days. These efforts proved the feasibility of saturation diving; subsequent projects would explore what this technique could accomplish for defence, science and industry.

Second-generation habitat projects Sealab II and Conshelf III took place oceans apart at the same time in summer 1965. Although Sealab I had been tested in the warm waters off Bermuda, the second American habitat was installed just off the end of the research pier at the Scripps Institution of Oceanography in LaJolla, California, at 60 m (200 ft). Planners intended to test the ability of aquanauts to do useful work in conditions similar to those found in the chilly waters on continental shelves around the U.S. Teams of ten divers spent two weeks each living in the habitat, but one aquanaut, the former astronaut Scott Carpenter, remained for four weeks. His participation capitalized on public interest in space exploration. A well-publicized radio link between Sealab and the Gemini space capsule allowed Carpenter to greet his former colleague Gordon Cooper, calling attention to the parallels of sea to space.

The project included studies of the human responses to working under stress – of interest to the government for its applications to defence and space exploration – as well as the testing and development of equipment and, taking advantage of the expertise at Scripps, geological and ecological studies of the sea floor and marine fauna. Tuffy, a Navy-trained dolphin, carried messages and delivered tools to aquanauts working at stations away from the habitat. Organizers eager to

Astronaut-turned-aquanaut Scott Carpenter atop Sealab II in 1965 before the habitat was lowered into the waters off La Jolla, California, where Carpenter spent four weeks on the sea floor.

prove the value of saturation diving to industry were most proud of tests related to salvage, such as testing a foam used to help raise a downed aircraft, deploying a collapsible salvage pontoon and operating a variety of power tools. They reported with great satisfaction that 28 aquanauts spent 450 man-days on the sea floor, completing more than 400 hours of useful work under adverse conditions and demonstrating the utility of habitats as a base of operations for commercial work.

Ten days before the end of the Sealab II mission, the American aquanauts spoke to Cousteau and members of his crew living in Conshelf III, installed east of Monaco in the Mediterranean Sea near Cap Ferrat lighthouse. The habitat's depth of 100 m (328 ft) was selected in response to new discoveries of oil deposits beneath the sea floor in the Gulf of Mexico, off the coast of California and in the Persian Gulf, fuelling a desire by oil companies to explore the feasibility of divers working at depths greater than 90 m (300 ft). The French 'oceanauts', as Costeau called them, lived for 22 days inside Conshelf III's two-storey, spherical habitat and, as before, used the diving saucer as a mobile base for tasks associated with installing and maintaining undersea oil-drilling equipment using a mock wellhead. Oil executives watching via closed-circuit television in Paris witnessed divers changing a 180-kg (400-lb) valve in 45 minutes, a promising demonstration of the utility of saturation diving for undersea oil exploitation.

Ocean boosters anticipated a new industry evolving around the capacity for saturation diving from sea-floor habitats. Many habitats were built and operated after 1965, including Hydrolab constructed by Perry Submarine Builders, Inc., and Aegir by Makai Range, Inc., a company founded by the dynamic scientist-turned-entrepreneur Taylor (Tap) A. Pryor. However, after Conshelf III, the oil industry parted ways from Cousteau and other habitat experiments, preferring to develop technology in-house and choosing the relatively safer option of having saturated divers live under pressure in surface chambers rather than in sea-floor habitats. Taylor Diving and Salvage, with funds from a buyout by Halliburton, the parent company of Brown & Root, built a research and training facility at Belle Chasse, Louisiana, in 1969 that could simulate dives to 300 m (1,000 ft).

Governments, not private investors or entrepreneurs, funded the majority of new habitats in the late 1960s. Well-known examples include Sealab III (built but not used) and Tektite I and II by the U.S., Chernomor by the Soviet Union, Seatopia by Japan and Helgoland by Germany. In their brief heyday in the late 1960s and early 1970s, habitats were designed to conduct scientific research and test engineering, no longer to serve as a base for human divers to execute industrial work.

THE ANTICIPATED OCEAN technology industrial sector had been expected to involve much more than just habitats, and it ended up raising questions about the ownership of the ocean that contributed to the erosion of freedom of the seas. In 1966 the President's Science Advisory Committee predicted, 'American industry may well be poised on the edge of what could, during the next 10 to 20 years, become a major, profitable advance into the marine environment.'[12] There were over a dozen new companies, often spin-offs of established defence or aerospace firms, which invested in ocean technology. A 1966 Lockheed advertisement in the journal *Science* introduced its new commitment to the sea with the question, 'Where can man go . . . in R & D?' and answered its own question in language laced with romanticism: 'To distant planets, to land-vehicles of the 1970s, and to a region far beyond the grasp of man today – the ocean bottom. Lockheed's major Research & Development programs reach from deepest space to the ocean deep.'[13] In 1965 Lockheed announced a new ocean science research facility; at the same time, other companies, including North American, Aerojet General and General Dynamics, among others, had similar facilities recently or nearly completed.

While only a few ocean technology firms built habitats, virtually all designed and constructed small submersibles intended for research and various industrial purposes. Examples included Ocean Dynamics, created by General Dynamics to build a series of small submersibles, *Star I, II* and *III*, for in-house ocean exploration and also for lease to other users. Likewise, Lockheed built *Deep Quest*, and Westinghouse, partnering with Cousteau, constructed *Deepstar 4000*. The resulting fleet of small submersibles had longer useful lives than habitats did. A few, such as Woods Hole's famous *Alvin* or the U.S. Navy's NR-1, a small nuclear submersible, had their working lives extended several times over. But these designs were never mass-produced for industry as anticipated, and most were not replaced when retired.

The ocean technology industrial sector expected defence and industrial uses of its hardware and expertise, the most eagerly awaited of which was probably the mining of manganese nodules from the deep ocean floor. Manganese nodules are rounded balls composed of minerals such as manganese, cobalt, zirconium and copper that often accrete around something like a shark's tooth. A great deal of excitement surrounded discussions in the 1960s of the economic potential of this resource, considered by many to be potentially the most valuable in the sea. Some boosters even touted it as 'renewable' because nodules are continually forming on the sea floor. The Chief Oceanographer of the U.S. Coast & Geodetic Survey suggested the possibility of establishing 'metal farms' once scientists understood the conditions that promoted accretion, extending the dream of farming the sea even to its non-living resources.

Mining the sea was not only a matter of profit but an urgent issue of national interest. The mineral wealth of the sea, in particular manganese nodules, drew attention to the deep-sea floor and to the lack of international agreement about ownership of its resources. As one journalist put it, leaning on the metaphor of the perceived lawlessness of the western frontier, 'Down into the cold, hostile depths of the sea, man and his technology are moving rapidly – but law to regulate this frontier is sadly lacking.'[14] Clarke's novel *The Deep Range* reflected the widely held belief that utilizing ocean resources would hinge on, or perhaps usher in, the end of nations and an era of international governance and collective ownership of resources. Without robust international law, or a single world government, global-scale ranching of whales or harvesting of plankton would be difficult or impossible.

Internationalist ideals competed with the realities of capitalist industry, for which control of resources was a precondition of the significant investment required for novel oceanic enterprises. The possible military uses of the seabed and its non-living resources prompted arguments for national control of parts of the ocean off the coasts of developed nations. In this era of unbridled technological optimism, the issue of political control of the ocean – particularly the sea floor and the seabed – appeared to be the only serious impediment to unleashing new industries. 'New legal concepts will have to be developed for the sharing of resources in and under the sea before we can colonize the oceans,' warned John Bardach, author of *Harvest of the Sea* (1968).[15]

Most of the new uses envisioned for the sea in the 1950s and 1960s involved the ocean's third dimension. The embrace of the frontier metaphor by ocean boosters drew attention to the importance of the underwater realm and the seabed in the 'Law of the Sea' process that unfolded from the late 1950s through the 1970s. By the nineteenth century, and strongly so after the Napoleonic Wars, Britain enforced both freedom of the seas for trade and navigation and a 5-km (3-mi.) territorial sea. Attempts by the League of Nations to codify international law of the sea in 1930 failed because small coastal states hoped to protect fisheries beyond 5 km, but the high seas continued to be accepted as free and open, and no significant territorial claims to the ocean were made until 1945. The new direction, after the Second World War, was prompted by the novel prospect of undersea oil resources, as well as the traditional resource of fish.

Immediately after the war's end, in September 1945, u.s. President Harry S. Truman issued two proclamations: first that the u.s. would regulate fisheries along its coasts, although these zones would remain high seas with no restrictions on navigation, and, second, that the u.s. claimed jurisdiction over the resources of its continental shelf extending offshore from its coastline. The Truman Proclamation set off what one scholar called 'the great sea rush of the 20th century'.[16] Nothing in existing international law prevented such a grab, which called attention to potential oceanic resources, including those other than fish, hidden in the depths. In response, a number of nations made similar claims to adjacent continental shelves or, in the case of Chile, Peru and Ecuador, a 320-km (200-mi.) zone extending beyond the bounds of their narrow continental shelves. Responding to these claims, the United Nations worked to codify international law related to the sea. In 1958, 86 nations convened at the first UN Conference on the Law of the Sea (UNCLOS) and adopted four conventions addressing legal regimes for territorial seas, the continental shelf, the high seas beyond and the living resources of the high seas.

The conference closed with several loose ends, which a follow-up conference in 1960 also failed to resolve. These included the breadth of the territorial sea, ambiguities in drawing baselines that determined the outward boundaries of national jurisdiction off a nation's coasts, and – most crucially for prospective ocean industry – an unsettled definition of the continental shelf. Not all nations accepted the conventions adopted in 1958. The objectors, mostly newly independent Asian, African and Latin American countries, were joined by more new nations whose defence of their interests in controlling the waters off their coasts contributed to the legal impasse with the major maritime powers which preferred continuing the policy of free seas. The third UNCLOS did not meet until 1973, so these important questions went unsettled for the entire decade of the 1960s while the ocean boosters cultivated the frontier metaphor and tried to implement the visions they had for human control of the undersea world.

The legal regime that ultimately resulted was constructed to a significant degree around the anticipated new uses for the ocean and its depths, although traditional military and fisheries concerns, of course, also shaped the outcome. The oceans served as a central arena for the Cold War, prompting the navies of powerful nations to prefer narrow

Delegates at the 1958 United Nations Conference on the Law of the Sea.

territorial seas to maximize mobility while preserving most of the ocean as free high seas. Distant water fishers resisted changes to the traditional freedom of the seas, but coastal nations, especially those without fleets of factory fishing vessels, cried foul, objecting to the idea of far-away nations catching fish from their adjacent waters. Oil companies, and others impressed with the fabled potential wealth hidden in the sea's depths, wanted national ownership of seabed and waters above the continental shelves, including beyond the 320-km (200-mi.) Exclusive Economic Zones (EEZS) which ultimately resulted from the Law of the

Sea process. EEZs were embraced as a compromise, allowing innocent passage for civilian vessels and transit passage for military ships, yet reserving the living and non-living resources of coastal waters and the sea floor for the adjacent nations.

In 1967 at the height of optimism about capitalizing on the ocean's resources, the Maltese ambassador to the UN, Arvid Pardo, proposed the concept of the high seas as the 'common heritage of mankind'. To widespread approval, he laid out a vision that the ocean's high seas resources should be understood as belonging to all people on Earth. He argued that sharing the wealth derived from the sea would address hunger and poverty in developing nations. This strategy of using the Earth's hitherto unclaimed and unreachable resources to solve profound and pressing global social and economic problems appealed to many people from many nations. Ocean industrialists did not welcome the idea of international governance, preferring the predictability of national control. Most observers, however, recognized that some parts of the ocean must remain international. The famous U.S. oceanographer Roger Revelle sharply criticized the idea of divvying up the ocean: 'The long-run consequences of such a division of the ocean into national territories are appalling to contemplate. They would constitute a *reductio ad absurdum* of the concept of nation-states.'[17] The reservation of a high seas area beyond national EEZs was the legacy of the combination of the long tradition of freedom of the seas and Pardo's vision, but an under-appreciated contributing factor was the incredible optimism about the prospect of drawing yet-untapped wealth from the sea and using it to equalize an unequal world.

THE OCEAN CAME TO BE viewed in the second half of the twentieth century through the cultural prism of the 'frontier'. The explosive growth of the science of oceanography, a response to the novelty of submarine warfare, transformed the ocean into a world system that scientists attempted to understand as such. In the Second World War and subsequent Cold War, the ocean's depths took on geopolitical importance as never before. Nations employed fishing for valuable species such as tuna, salmon and whales to exert territorial claims that, while not formally territorial, involved the assertion of national power. Competition for real and perceived marine resources fuelled a series of debates and unilateral national actions that resulted in the global enclosure of much of the sea through Exclusive Economic Zones. The frontier metaphor encouraged the assumption that the sea's resources were essentially limitless and the expectation that engineering and technology would enable firm human control of the ocean and its depths.

The perception that the ocean offered endless mineral and food resources promoted the idea of the ocean, including its third dimension, as a space available for work. Increasingly, its depths were embraced as an arena for play as well, drawing the undersea realm into popular culture in a way that echoed the nineteenth-century discovery of the sea. As the next chapter chronicles, recreation emerged as a major use of the sea with the development of the cruise industry, the enthusiastic embrace of scuba and the appearance of whale watching and other activities that now together comprise a major economic sector very different from the futuristic work at and in the ocean envisioned in the 1960s. The frontier metaphor for the ocean and its depths yielded, haltingly over many decades, to a re-imaging of the sea as a wilderness – a place outside human society but in need of protection.

⌐ SEVEN ⌐

Accessible Ocean

No one can build a moon rocket and blast off from his own
backyard, but almost anyone who wishes may still take part
in the exploration of inner space.
– Alexander McKee, *Farming the Sea* (1969)

WHALE RANCHING, PLANKTON burgers, undersea oil refineries and submarine cargo transport – all dreams of 1960s ocean boosters – did not come to pass, and neither did the ocean's depths remain exclusively a place for work. Instead, the accessibility of the undersea world led to the refashioning of the ocean as a place for play and an environment suddenly visible to people through personal encounters and popular culture media, especially thanks to underwater filming. The undersea world entered the realm of mass culture more thoroughly and completely than during the nineteenth-century discovery of the ocean. Through recreation and popular media, a privileged role for science remained as a means for knowing the ocean. 'Everyone who goes underwater becomes a scientist,' declared science fiction author Arthur C. Clarke in 1960, at the height of his ocean boosterism.[1] In contrast to outer space, the undersea realm proved accessible in 1954 to anyone willing and able to spend $10 on skin diving equipment – snorkel, googles, and fins – or $160 for an Aqualung, the first self-contained breathing apparatus marketed commercially.[2] A colour television set at the time cost over $1,000. Space exploration remained a future dream; the Russians launched Sputnik in 1957 and sent Yuri Gagarin, the first person in space, into orbit in 1961. The American Mercury and Apollo space programmes, touched off by Sputnik, put Neil Armstrong and Buzz Aldrin on the Moon at the end of the decade, in 1969. As the superpowers spent billions to enable a few individuals to reach space, thousands of people started exploring the ocean's third dimension personally, and millions more through books and film. Their collective experiences initiated a revolutionary reorganization of the human relationship with the sea.

BEFORE THE UNDERSEA REALM could be visited by people other than expert divers using diving bells, hardhat suits or traditional techniques that didn't rely on specialized equipment, it was envisioned imaginatively and at times created artificially. Decades before the aquarium's invention in the mid-nineteenth century, the geologist Henry De la Beche employed fossil evidence to create the first pictorial representations of deep time, including an underwater view of plesiosaurs, ichthyosaurs and other marine creatures from the Devonian era attacking each other. He may have been inspired by his experience of diving off the Dorset coast. Actual aquaria provided the first glimpses of animals and plants in their underwater realm. While small home tanks brought the ocean into private homes, large public aquaria were founded in the second half of the nineteenth century, first in London and in the U.S., and then in a number of European countries and in Japan. These popular attractions introduced sea creatures and underwater scenes first to residents and visitors of coastal and capital cities, and, by the first decades of the twentieth century, to those who lived far inland. Chicago's Shedd Aquarium, for example, built in 1929, remained the largest aquarium in the U.S. through the century.

In 1938 a novel facility opened in St Augustine, Florida, that paired the display of marine animals with the burgeoning film industry. Marine Studio founders designed tanks to give visitors screen-like views into the underwater realm. The spectacle of the sharks, manta rays and dolphins displayed there immediately overshadowed aquarium exhibits that lacked such large and dramatic creatures, and tourists arrived by the thousands. But attracting visitors was only part of the mission of this new institution; founders designed its tanks for use by film-makers to satisfy the growing market for underwater scenes in movies and other media.

The first underwater motion picture was not filmed in tanks but directly in the waters off the Bahamas, and it premiered more than two decades earlier. John Ernest Williamson, an avid reader of Jules Verne and Victor Hugo, developed an invention of his sea captain father, used for ship salvage and underwater repair: a large tube of concentric rings that could be extended and collapsed like an accordion, enabling a person to work beneath a boat yet remain dry and breathe air pumped down to him. To the end of the tube, Williamson attached his 'Photosphere', a spherical observation platform from which a photographer could film activity in the water illuminated by lights mounted on the boat that deployed the apparatus. He was commissioned to film the underwater scenes for the 1916 movie *Twenty Thousand Leagues under the Sea*, which achieved great success at the box office, no doubt appealing to an audience closely following news of submarine warfare. Williamson devoted his career thereafter to making motion pictures featuring mermaids and shipwrecks, sunken treasure or sea monsters, chronicling his work in his 1936 autobiography, *20 Years under the Sea*. He also worked with scientists at the American Museum of Natural History in New York and the Field Museum in Chicago to collect specimens of coral and fish for reef dioramas. While Williamson continued to film

Underwater photograph of an actress taken with John Williamson's Photosphere.

in the sea throughout his career, most Hollywood directors preferred the more controllable environment of tanks, creating a ready market for Marine Studios and similar facilities.

At about the same time that Marine Studios opened, an American former pilot, Guy Gilpatric, invited readers to explore the underwater world for themselves by introducing the new hobby of skin diving. Coastal cultures around the world have, for thousands of years, used diving to gather food and valuable trade items and during warfare. Ancient Greek divers fished for sponges, constructed underwater defences, sabotaged enemy ships and salvaged treasure and cannon from shipwrecks. Stories about Alexander the Great exploring the sea floor using a diving bell, some of which chronicle a three-day visit under the sea, may be based on his use of divers to remove underwater obstacles during a siege, work that he may have observed personally by using a diving bell. During the nineteenth century, a handful of scientists made single, or perhaps several, dives, but diving otherwise remained mainly a military or economic activity for tasks such as laying lighthouse or bridge foundations, destroying enemy ships and marine salvage. Gilpatric invented skin diving in the warm Mediterranean waters of the French Riviera, first modifying goggles used by aeroplane pilots to make them waterproof, and then fashioning a spear that could be used underwater to catch fish. A writer who later authored a number of books and a series of short stories chronicling his famous character, the Scottish ship engineer Mr Glencannon, Gilpatric penned a humorous guidebook, *The Compleat Goggler* (1934), to introduce his new sport of 'goggling'.

Like traditional divers in other cultures, most early skin divers hunted fish and other seafood. In the U.S., the Depression prompted divers to take up spearfishing, even in chilly California coastal waters, to feed their families. Yet the diving craze initiated by goggling diverged from the practice of using diving as a tool for work. Most twentieth-century skin divers entered the sea for recreation, not for subsistence, warfare or paid labour. Before diving completely opened to mass participation, though, its use during the Second World War contributed new technologies and inspiration to novice divers who admired the exploits of naval frogmen. A French naval officer serving in Toulon, Jacques Cousteau, borrowed underwater goggles from a friend shortly after Gilpatric's book was published. With a small group of military comrades, he tinkered with diving gear before and during the Second World War.

Mass participation in diving depended on technological innovations that transformed military diving equipment to make operating underwater easier and safer. When Cousteau partnered with engineer Émile Gagnan, they developed a valve that could regulate a diver's air supply and patented the resulting Scaphandre Autonome, or Aqualung, in 1943. Initially the pair assumed their diving gear would mainly appeal for military or other kinds of work, but when Gagnan emigrated to Canada after the war, they decided to market the Aqualung in North America.

Cousteau's Aqualung became the first self-contained breathing apparatus available for purchase by ordinary people. Technological access to the undersea realm not only spawned ambitions to extract wealth from the sea but equally transformed the ocean's third dimension into a place of recreation. From the moment scuba technology appeared on store shelves in the late 1940s, many people took the plunge. Diving instruction manuals followed Clarke's cue, framing undersea sport as discovery and urging its adepts to eschew mere adventure for scientific exploration. By 1970, the cookbook *Bottoms Up Cookery* addressed divers graduating from beginner status and urged them to study the habits of sea creatures in order to hunt them successfully. Its cover illustration shows a diver examining the sea floor with a magnifying glass, testifying to the interpretation of science as a meaningful way to engage with the depths, even if only to capture dinner.

Among the earliest to adopt the first Aqualung devices imported into the U.S. in 1949 were two graduate students at the University of California and also several members of the San Diego Bottom Scratchers diving club, founded in 1933 as a club for breath-holding underwater

Cover of *Popular Science* magazine from 1953: the diving craze drew men, women and children into the underwater world.

hunters. In the 1950s it served as home base to divers who pioneered the use of scuba for underwater photography, archaeology and marine science.

The well-established Bottom Scratchers club shared the underwater realm by the mid-1950s with over two hundred other dive clubs in the U.S. and almost fifty elsewhere in the world. Pun-filled names for clubs conjure up the social nature of organizations that formed to educate divers, run spearfishing contests and connect divers with institutions that needed their expertise, such as local police stations for search and rescue operations. Examples include 'Davy Jones Raiders' in Lynwood, the 'Kelptomaniacs' of Los Angeles and the 'Sons of Beaches' of Long Beach, all three in California, whose shores served as scuba's incubator, according to diving historian Eric Hanauer.

Although the West Coast hosted the majority of clubs, many also formed on the East Coast, including the erudite-sounding

The original members of the San Diego Bottom Scratchers diving club in 1939, standing in the La Jolla Coves area. Left to right: Glen Orr, Jack Corbaly, Ben Stone, Bill Batzloff and Jack Prodanovitch.

'Anthro-Piscatoral Society of Connecticut' in New Haven, along with clubs in New York and Florida and a handful in most of the New England states and New Jersey. Places in between the Atlantic and Pacific coasts had clubs, including two in New Orleans, four in Chicago and others in Great Lakes states such as Michigan and Wisconsin, but there was also the landlocked 'Arizona Desert Divers Club' of Phoenix. Although most of the world's dive clubs in 1956 were in the u.s. or in American territories, there were 24 in Italy; 6 in Australia; 3 each in France and Mexico; 2 each in Canada, South Africa and England; and 1 each in Japan, Algeria and Curaçao. Divers from eighteen countries met in 1958 in Brussels, Belgium, to establish an international organization, the Confédération Mondiale des Activités Subaquatiques, which opened its office in Monaco.

The Underwater Society of America formed one year after its international counterpart. Its headquarters location, in Champaign, Illinois, emphasizes that the category of 'undersea' was hardly limited to the ocean. As Bill Barada, author of a popular diving manual, explained, 'Every new vista encountered invites exploration, and not only in the sea. An abandoned quarry, small lake or quiet river offers limitless opportunity for adventure. Almost every body deep enough for swimming is being probed by skin divers.'[3]

The number of divers grew even faster than the number of clubs. In 1949, when *National Geographic* published an article on the San Diego Bottom Scratchers, southern California was home to about 8,000 divers. By 1951 enthusiasts had their own publication, *Skin Diver* magazine, which began as the kitchen-table production of two members of the Compton, California, 'Dolphins' dive club. Subscribers in 1957 received a free reprint of Gilpatric's *Compleat Goggler*. *Skin Diver* soon became the largest diving magazine, so successful that a publishing company purchased it in 1963. By 1965 one popular book on diving estimated that more than six million Americans were divers.

Already in the early years of recreational scuba diving, women and even children joined men underwater. Although military diving was

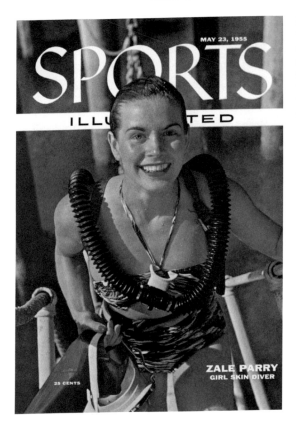

Zale Parry was featured on the cover of *Sports Illustrated* magazine after she became the first woman to dive below 60 m (200 ft).

exclusively a job for men, 10 per cent of the 50,000 members of the U.S. National Frogman Club in 1955 were women. Diving instruction manuals featured photographs of women and children learning to dive, as if to reassure readers that the sport was safe for all. The introduction of dive training and wetsuits transformed the sport and opened it to novices, to men and women, and to young and old. A 1959 AMF-Voit sporting equipment advertisement in *Life* magazine showed boys from Explorer Post 3 in inland Shelby, Ohio, helping one another with scuba gear as they prepared to demonstrate their newly acquired diving skills in Walton Lake.[4] Another *Life* feature two years later used the caption, 'Tucking in a Tadpole' to describe a photograph of a father helping his four-year-old son adjust his wetsuit and scuba gear before entering

Lake Winnipesaukee, New Hampshire, where the father worked as a diving instructor.[5]

Women started some of the earliest scuba clubs and numbered among the first certified scuba instructors, underwater film stars, and stunt doubles. Wives and girlfriends of the Bottom Scratchers started the first women's dive club, the Sea Nymphs. One member, Zale Parry, began diving using air left in the tanks of her date and his diving buddies when they got too cold to dive. She reminisced, 'It was a long time before I dived a fresh bottle. I probably made the most free ascents in the world; they were just part of the sport to me.'[6] In 1954 Parry broke the women's depth record, diving below 60 m (200 ft) in an attempt sponsored by diving companies that wished to promote the sport to the general public, especially to women. *Sports Illustrated* featured her achievement on its cover. The third woman to be certified as a scuba instructor in the U.S., Parry made a career in the film industry starting with stunt work and a few starring roles in the television series *Sea Hunt*.

PARRY'S CAREER EXEMPLIFIES how women's participation in diving, and also in scientific and popular writing about the ocean, helped transform the sea into an accessible environment. As a visible figure in the emergent diving community, she served as a model for prospective women divers and, through her involvement in scuba instruction and television work, she actively promoted undersea activities as exciting and accessible. The scientist-diver Eugenie Clark likewise demonstrated through her work, her day-to-day life and her writing that ordinary people could explore the ocean. Her books *Lady with a Spear* (1953) and *The Lady and the Sharks* (1969) chronicled her innovative application of freediving (holding one's breath), and later scuba, to her research on fish, invertebrates and sharks. For Clark, diving became a family activity. Her husband learned to dive on their honeymoon, and she taught her four children, as did another well-known mother, botanist, ocean explorer and writer who began her career several decades

Eugenie Clark, a marine biologist who used diving to observe fish, collect specimens and, later in her career, study shark behaviour.

later: Sylvia Earle. Both Clark, whose husband was a physician, and Earle dove while pregnant, a practice that seemed quite natural to their doctors and other experts at the time. Through her popular books, Clark presented diving as an activity that enabled ordinary people to know the ocean, assuring readers, 'Once you are familiar with the sea's dangers and know where to expect them and how to avoid them, you can roam with safety and assurance in a world of wonder that otherwise you will never really know.'[7]

Clark's writings and those of Rachel Carson introduced the reading public to scientific understandings of the ocean newly available after the Second World War. The two authors shared the same literary agent, and both admired William Beebe, the explorer who wrote about the observations he made of deep-ocean creatures at depths of up to 900 m (3,000 ft) in the early 1930s while riding in his bathysphere, a spherical hull with a porthole that was lowered into the sea on a steel cable. At a time when underwater films and public aquaria attracted public interest, Beebe's dives to abyssal depths received enthusiastic coverage in the media. Beebe took steps to present his work as science, not as adventurous record breaking, even when he broadcast a 1932 dive on the radio and including the occasion when he and his partner Otis Barton were lowered in the bathysphere to the depth that provided the title for his popular book *Half Mile Down* (1934). Beebe's writings about deep-sea fish and his experiences in the dark abyss emphasized the insignificance of humans in the vastness of the ocean. In stressing the danger from pressure, the awe and desolation of the darkness, and the strangeness of the colours of phosphorescent creatures, Beebe depicted the ocean's depths as sublime, mysterious and beautiful, a tradition that Carson's work extended.

Carson's critically acclaimed book *The Sea Around Us* (1951) was her second about the ocean. As a student, she had wrestled with a

William Beebe standing to the left of the Bathysphere, with Otis Barton Jr, c. 1930–1932.

decision about whether to pursue writing or science, believing them incompatible. Choosing science, Carson began a career as a civil servant in the Bureau of Fisheries (later the Fish and Wildlife Service), but began writing popular articles outside work to earn extra money. Her first book, *Under the Sea Wind* (1941), traced the journeys of Scomber the mackerel, Anguilla the eel and a black skimmer, a seabird in the gull family, named Rynchops. Careful to avoid anthropomorphism, Carson portrayed the ocean world of shore, open sea and abyss from the perspective of the animals, with no narrator and humans appearing only as predators or destroyers. Readers saw the interdependence of all the sea's creatures, gaining an ecological understanding of the sea through Carson's poetic and imaginative evocation of the wonder of nature's eternal cycles. The book came out a month before the Japanese attack on Pearl Harbor and, although it received enthusiastic reviews, sales were disappointing, no doubt due to national preoccupation with war.

During the war, government-sponsored scientists produced a massive amount of new research on the ocean in support of submarine and air warfare, amphibious landings and surface naval activities generally. After the war, Carson's job included editing government reports on newly declassified material, providing her with a front-row seat to view the emerging oceanographic knowledge. Carson scooped other writers by publishing the first popular book revealing the new knowledge gained about the ocean environment by wartime science. *The Sea Around Us* won numerous awards, including the National Book Award and the prestigious John Burroughs Medal awarded by the American Museum of Natural History. It sold 250,000 copies within a year, and was translated into 32 languages, testifying to the excitement surrounding new ocean discoveries. The list of leading scientists she thanked in her acknowledgements demonstrates the extensive professional and personal network she drew upon in this and her other popular science writing. The success of *The Sea Around Us* enabled Carson to quit her job at the Fish and Wildlife Service and devote herself to writing.

Although Carson was neither a sailor nor a diver, she took advantage of opportunities while working on her book to accompany the Fish Commission research vessel *Albatross III* on a cruise and to dive with the director of the University of Miami's new marine station. The immensity of the world of water she saw from the deck of *Albatross* reminded her of the vast perspective of geological time, while the sight of nets hauled from the depths full of fascinating creatures made an unforgettable impression. Off Miami, rough conditions at sea limited her to a brief, nervous time underwater in a diving helmet, clinging to a ladder, but she carried away with her the memory of how the surface looked from below and believed that the experience contributed vitally to her ability to write effectively about the ocean.

In a fellowship application for time away from her job to write, Carson explained her goal for the book as 'an imaginative searching out of what is humanly interesting and significant in the life history of the Earth's ocean.'[8] As others of her time were, Carson was impressed by

Rachel Carson exploring a tidepool on the Atlantic coast with Bob Hines in 1952, the year after she published her acclaimed work *The Sea Around Us*.

the promise of 'the wealth of the salt seas', but she wanted to convey instead the beauty and mystery of the ocean's role in the Earth's history and in fostering life.[9] She did not shrink from the destruction wrought by humans, and hoped to inspire appreciation for the sea and commitment to seashore preservation.

Women did not, of course, pen the only popular books about the ocean, as Beebe's work attests. *The Sea Around Us* shared the U.S. bestseller list of 1951 with other titles that suggest readers' curiosity about the sea, including Thor Heyerdahl's *Kon-Tiki* and Herman Wouk's *Caine Mutiny*, among others. But the extensive involvement of women and children – non-experts in general – in exploring the ocean imaginatively through books and bodily through recreation contributed to the domestication of the sea.

TO THE FIRST GENERATION of divers in the 1950s, before the ocean underwent domestication, the depths were filled with danger. Clark began studying sharks, creatures that had long filled mariners with dread and that, during the Second World War, inspired determined but ultimately unsuccessful efforts by researchers to create shark repellents to protect pilots downed in the ocean. Octopuses, sharks, moray eels and other menacing creatures seemed threatening to early divers; that these divers entered the sea to hunt perhaps fuelled this antagonistic vision. Soon, though, spear-wielding adventurers were joined underwater by divers who preferred to stalk with cameras or by people who simply wished to observe and experience the undersea world. Experience redefined the giant octopus from monster to recluse and taught divers how to operate around creatures like moray eels without incident. Sharks, divers learned, might be unpredictable, but that was less frightening than the vision of them as bloodthirsty monsters that attacked humans on sight.

Within the first decade of recreational diving, commentators concurred that the greatest danger was posed not by the sea but rather by the human element. Divers who pushed the limits of the existing equipment and techniques exposed themselves to 'rapture of the deep', or nitrogen narcosis, caused by going too deep while breathing pure oxygen. The implication was that better technology and training could neutralize danger, and, indeed, the adoption of mixed gas breathing reduced the danger of narcosis. By the mid-1960s, diving manuals concurred with Clark that the undersea environment itself, at least to proven recreational diving depth limits, was 'a safe and friendly place'.[10] Despite such reassurances, the plots of televisions programmes such as *Sea Hunt* turned on terrifying encounters with sharks and epic personal struggles against rapture of the deep.

Movies, television shows and books brought the undersea realm into people's minds and also into their living rooms, schools and local movie theatres, offering readers and viewers vicarious experience and probably spurring some to try diving for themselves. One clear signal of the shift away from the undersea as workspace emerged with Cousteau's opportunity to sell film footage from his Conshelf III habitat experiment to the American network CBS for a one-hour television special. With the loss of financial support from oil companies, Cousteau benefited from burgeoning interest in underwater action by the film and television industries. His 1953 book *Silent World* inspired an Academy Award-winning film released three years later, which was soon joined by other media that peered beneath the waves. The 1954 version of Jules Verne's *20,000 Leagues under the Sea* drew crowds and itself became a classic. In the 1955 monster movie, *It Came from Beneath the Sea,* the navy submarine commander protagonist, who had been tracking the gigantic, threatening octopus that had been irradiated by hydrogen bomb testing, donned an Aqualung to fight the octopus on the ocean floor at close quarters in a plot that seems influenced by *Godzilla*, released the previous year. Also in 1955, the eccentric industrialist billionaire Howard Hughes produced *Underwater!* starring Jane Russell and promoting the new scuba technology. Beginning in 1958, *Sea Hunt* gained enough popularity to inspire the similar 1960 series, *Assignment Underwater.*

Both ended in 1961, the year Irwin Allen's film *Voyage to the Bottom of the Sea* premiered. Its namesake television series opened in 1964 and ran for four years. In 1953 Allen had produced a documentary version of Carson's *The Sea Around Us* (1953) that angered her with its inaccuracies and anthropomorphism and garnered criticism but won an Academy Award in the documentary category. In the same year that Conshelf III divers moved into their new habitat, the James Bond film *Thunderball* (1965) thrilled viewers with lengthy underwater action scenes involving both sharks and bad guys.

Scuba also featured in advertisements for a wide variety of products, especially in order to associate these with outdoor activities and adventure. Scuba was used to sell 7-Up and Pepsi as well as alcoholic beverages, including Ballantine beer and Canadian Club Whiskey. Coppertone tanning lotion was perhaps a natural, but Tampax was also touted for active woman divers. And virtually every car manufacturer, including Austin Healey, Pontiac, Chrysler and Mercury, deployed scuba to promote various car models. Advertising reinforced the messages conveyed by films and television shows, as well as diving instruction manuals and popular science books, that the underwater zone was not a dangerous and unknown realm but a mysterious and beautiful world that invited personal exploration.

Such a world enticed tourists from among the wealthy and middle class Westerners during the prosperous post-war decades. Surfing had a longer history than scuba, originating as part of ancient Polynesian culture but re-emerging in the early twentieth century in Hawaii and spreading to Australia and California. The popular movie *Gidget* (1959), about the real-life young California surfer Kathy Kohner-Zuckerman, marked the emergence of surfing as a popular sport that evolved its own soundtrack via music such as the Beach Boys' 1962 album *Surfin' Safari*. Alongside experiments with saturation diving to achieve industrial uses of the undersea realm, dreams emerged in which the benign underwater world would accommodate tourists at submerged hotels built on coral reefs. In 1960 futurists predicted with confidence facilities where guests could wave from giant observation windows to friends embarking on guided excursions far from the hotel using futuristic jet propulsion devices. The Futurama exhibit prepared by General Motors at the 1964 New York World's Fair touted the ocean's depths as the newest vacation playground, among examples of people mastering remote and challenging environments that also included Antarctica and outer space.

While undersea hotels were not forthcoming, this impulse to promote ocean tourism found success on the high seas with the emergence of the modern cruise industry in the late 1960s after jetliners supplanted ocean liners for transportation across oceans. The boost to the cruise industry by the popular television show *Love Boat*, which premiered in 1977, echoed the role of popular culture in the emergence of scuba as means for knowing the ocean.

Although the Futurama vision celebrated the conquest of nature's extremes, the accessible undersea realm, stripped of danger, attracted new metaphors. If Clark and the legions of recreational scuba divers

Hotel Atlantis, at the Futurama exhibit at the 1964 New York World's Fair.

domesticated the ocean, Carson created a novel ocean-centrism in which the sea remained enigmatic but awesome and appealing as a result. About the ocean's third dimension, she admitted, 'even with all our modern instruments for probing and sampling the deep ocean, no one now can say that we shall ever resolve the last, the ultimate mysteries of the sea.'[11] Even as the frontier metaphor took hold, many writers and observers embraced the related conception of the ocean as wilderness, as historian Gary Kroll argues.[12] In 1967 *Life* magazine labelled Tap Pryor, the former marine biology graduate student who founded Hawaii's first oceanarium, Sea Life Park, a 'frontiersman of the sea.'[13] Alongside the public aquarium, he created the Oceanic Institute, whose mission of scientific research would, he hoped, be supported by profits from tourists attracted to the Park. His wife and partner in these related ventures, Karen Pryor, took charge of training dolphins. A *Reader's Digest* article the following year praised the combination of 'exciting entertainment with top-flight oceanographic research', referring to the couple as 'pioneers, whose wilderness is a wet one.'[14]

MARINE MAMMALS, PARTICULARLY dolphins, figured prominently in the conversion of the ocean from an industrial workplace to wilderness in need of human protection. The first step involved a decisive shift in perception of marine mammals away from swimming commodities, in the case of whales, or pests to the fishing industry, the reputation of dolphins. One institution particularly fostered the rehabilitation of dolphins into friendly and intelligent creatures. The same Marine Studios in St Augustine, Florida, that served the emerging underwater film industry combined science with spectacle. The institution's facilities were intended from the start to support research, and scientists arrived throughout the 1940s and 1950s to study the respiratory physiology of diving mammals. After the war, dolphin communication and echolocation became important areas of investigation. Visiting scientists supported spectacle by providing advice that improved water quality and the health of animals exhibited, but the dolphin shows quickly overshadowed other activities.

Although the earliest dolphin training derived from traditional animal training practices, promotional materials for Marine Studios touted the role of science, later fulfilled when B. F. Skinner's operant conditioning was applied to training these mammals. Flippy, known as the 'educated porpoise', garnered fame in the early 1950s, prompting the construction of a stadium with seating for 1,000 people to see his tricks. Raising a pennant to open his act, Flippy jumped through a paper-covered hoop, rang his own dinner bell, performed barrel rolls and finished by towing a woman and small dog around his tank on a surfboard. Soon Marine Studios, renamed Marineland, opened a park in California and gained competition when Miami Seaquarium opened in 1954, while Sea World began operation in San Diego a decade later.

Visitors to the new parks also enjoyed seeing marine mammals in films. In 1946 the producers of *Bambi* created the Disney short film *The Whale Who Wanted to Sing at the Met*, featuring Nelson Eddy as Willie, the friendly and talented singing whale who dreams of performing opera for appreciative audiences. While the sailors who hear Willie are amazed, the famous opera impresario Professor Tetti-Tatti is unable to believe that a whale can sing. He concludes that Willie has swallowed an opera singer, and harpoons the whale in the hope of saving the musician inside. As *Bambi* did, the film included tragedy; unlike *Bambi*, it was the main character who died, but the narrator quickly deflected sorrow by reassuring the audience that Willie will sing in heaven. The 1963 film *Flipper* reveals similarly evolving impressions of dolphins. Flipper and Sandy, the boy who saves him from a spear injury, must find a way to convince Sandy's fisherman father that dolphins are not enemies but can be real friends.

Dolphin shows delighted audiences, who came to view these mammals as friendly and intelligent. Behind the scenes at these parks, dolphin research revealed aggressive sexual behaviours by male dolphins and at times involved invasive experiments on dolphin brains that killed many

Postcard showing feeding time for dolphins at Marine Studios in 1938.
The popularity of dolphin shows transformed the institution, renamed
Marineland, into a popular attraction in the 1950s and '60s.

experimental subjects. Discoveries about dolphins' ability to use sonar
to find objects underwater and sound to communicate with each other
attracted the military to cetacean research but also contributed to a
popular scientific view of dolphins and other marine mammals as highly
intelligent creatures with larger brains relative to their body size than
humans. Some futurists predicted that learning to communicate with
dolphins would prepare people for encounters with alien life. More down-
to-Earth observers cited cetacean intelligence as a reason to question
the continuing industrial hunting of whales.

In the early 1970s, experiments on dolphins by John Lilly that began
with Cold War funding devolved into highly questionable investigations
involving Lilly sharing LSD with dolphins, among other activities. His
work, including his book *Man and Dolphin* (1961), inspired a number of
writers and others to embrace the idea of cetaceans as highly intelligent,
even possibly uniquely capable of higher states of consciousness than
humans. The story 'The Voice of the Dolphin' (1961), by the Manhattan
Project physicist and critic of atomic weapons Leo Szilard, featured
dolphins guiding humans to nuclear disarmament. The activist Scott

McVay, who read Lilly's first book and also his book of 1969, *The Mind of the Dolphin*, urged his own readers to understand human survival as linked with that of whales. McVay teamed up with biologist Roger Payne to record and analyse humpback whale songs. The resulting album, *Songs of the Humpback Whale* (1970), translated the activists' verbal arguments into haunting music heard by millions of people who were encouraged to think of the sound as cries for help. A collection of essays and poems appearing in 1975, *Mind in the Waters*, presented whales as peaceful creatures who used their intellectual capacity to think and communicate rather than to manipulate tools and kill other beings. The emerging environmental movement seized upon whales as symbols of innocent and intelligent natural creatures in danger of eradication by a brutal industry.

One group of whales emerged as a bellwether for cetacean survival. The grey whales, which migrate perhaps the farthest of any mammal, give birth and nurse their calves in the sheltered lagoons of Baja California. Starting relatively late in the wooden whaling ship era, they found themselves the object of an intense fishery when an American vessel on its way to polar seas entered Magdalena Bay in 1845 and saw countless spouting whales. The crew soon learned why grey whales had earned the nickname 'devilfish', as the mother whales fought to protect their young and themselves. Soon whalers learned to harpoon the babies to lure the mothers into shallow water where it was easier to kill them. Lagoon whaling, practised in earnest in the late 1850s, reduced the grey whale population by about 90 per cent within a decade. The whales' vulnerability may have ultimately saved them. In recognition of the grey whale's extremely reduced numbers, they were protected by international regulation in 1938, almost half a century before the International Whaling Commission instituted the commercial whaling moratorium of 1985. In 1950, 10,000 visitors observed migrating grey whales at the Cabrillo National Monument in San Diego, California, and five years later whale watching from boats permitted closer views. By 1967, when environmentalist and author Wesley Marx wrote about them, grey whales were believed to number around 6,000, a modest recovery from what might have been a low point of 1,000 adult females. At present, the American Cetacean Society considers the current population of 19,000 to 23,000 individual grey whales, close to its original size, a 'remarkable recovery'.[15]

WESLEY MARX'S BOOK *The Frail Ocean* (1967) was another bellwether, signalling a new attitude towards the ocean in a decade of unbridled enthusiasm for treating the sea as a limitless frontier. While some environmentalists focused on whale killings, Marx embedded his worry about charismatic marine megafauna in a wider concern for the ocean. He called attention to pollution and how it degraded habitat. Grey whales, despite protection from exploitation, might not fare well, he warned, if the secluded lagoons where they retreated to give birth and nurse their young became filled with salt works, deep-water ports or other industries. Marx likewise noted that dredging or filling in estuaries often damaged crucial nurseries for valuable commercial species, and he called attention to the dangers of accumulation of mercury and DDT in fish and seabirds, as well as the threats posed by sewage, unexploded ordnance and other materials dumped at sea. Most of Marx's warnings related to coastal waters; offshore waters continued to be considered immune to the effects of pollution, while high-seas resources other than great whales still prompted advocacy for international cooperation to allocate them fairly rather than efforts to protect them.

One issue particularly directed the attention of environmentalists towards the sea. The 1967 wreck of the oil tanker *Torrey Canyon* 32 km (20 mi.) off the coastal resort area of Land's End in Cornwall, England, revealed a devastating danger posed by the lucrative and rapidly expanding oil industry. Within three days, oil slicks stretched over more than 250 km^2 (100 square mi.) of ocean and covered beaches first nearby and, in the following weeks, as far away as the French peninsula of Brittany. Just two years later, as Marx reported in subsequent printings of his book, the sensational blowout under an oil platform 10 km (6 mi.) off the coast of Santa Barbara, California, spread oil over 2,000 km² (800 square mi.) of coastal waters and 48 km (30 mi.) of beaches. Media images of dying, oil-soaked birds and sludge-covered beaches riveted the attention of a public increasingly concerned about the environment, and many observers credit this spill as the event that sparked national attention to environmental issues.

Despite the enthusiasm to protect whales and the outrage over oil spills that washed up on beaches, the ocean was edged out of mainstream environmental concerns in the flurry of environmental activity touched off by Rachel Carson's *Silent Spring* (1962) and other events. The fallout in the u.s. from Earth Day in 1970 and from inland disasters such as killing smogs and the 1969 burning of the Cuyahoga River focused attention on issues related to clean (fresh) water, clean air and other terrestrial matters. Ocean issues that might have generated concern, such as overfishing or ocean dumping, were interpreted in ways that deflected worry or action. Fish stock crashes prompted not fishing restrictions but efforts by scientists to help fishers find new populations and species to exploit. Marine chemists continued to believe the dictum that the solution to pollution is dilution, remaining convinced that the open ocean was a safe place to dump many kinds of waste.

Just as Carson's work on DDT rested on threats to human health, the dangers from eating poisoned seafood represented another exceptional case in the otherwise-terrestrial environmental movement. First diagnosed in 1956, the neurological disorder Minamata disease was determined to be caused by the consumption of fish contaminated with methyl mercury. In 1966 the International Council for the Exploration of the Sea formed its first marine environmental science research group, the Fisheries Improvement Committee, to address issues of marine pollution and the safety of shellfish consumption, although the original motive for the shellfish work related to aquaculture, not environmental issues. Oil spills, red tides and pollutants in coastal waters prompted scientific investigations and, in some cases, led to steps to prevent future

A blowout at a drilling platform 6 miles off Santa Barbara, California, on 28 January 1969 led to the largest oil spill in U.S. waters to that time.

accidents or to control the inflow of undesirable chemicals or nutrients. Mainstream environmental organizations were slow to focus on the ocean, with the noteworthy exception of Greenpeace and, later, Sea Shepherd, dedicated to ending the killing of whales. Older organizations such as the Sierra Club, the Audubon Society and the Wilderness Society remained terrestrial in orientation, as did the many new environmental groups that formed. The pressing environmental issues pursued between the first Earth Day and its twentieth anniversary resuscitation in 1990 included energy, air pollution, rainforest destruction, soil erosion, ozone depletion and acid rain – in short, mostly land-oriented challenges.

THE FORMAL ENVIRONMENTAL MOVEMENT of the 1960s rested on foundations as terrestrial as the late nineteenth-century tradition of conservation, in which governments took steps to preserve forests and manage their resources to ensure future use. In parallel, advocates influenced by romanticism argued for preserving wilderness to ensure that harried urbanites and future generations would have restorative, untouched wild areas to experience. An initial focus on forests and mountains expanded to include deserts, swamps, plains and other types of environments on land but did not extend to the ocean, which continued to be viewed as a timeless place, immune to human influence. While twentieth-century fisheries scientists and managers throughout Europe and North America advocated scientific conservation for the rational exploitation of fish stocks, the enterprise remained confident about the prospect of managing the marine system for consistent, maximal yields. No one emerged to argue for preserving areas of the ocean untouched, although the sea came to be seen as 'wilderness' in addition to 'frontier'. The frontier label no doubt contributed to the century-long delay in recognizing the ocean as needing, and deserving, the kinds of protection extended to land, air and fresh water. Perhaps the wilderness association planted new seeds that are just now bearing fruit.

Epilogue:
Ocean as Archive,
Sea as History

Time itself is like the sea, containing all that came before us,
sooner or later sweeping us away on its flood and washing
over and obliterating the traces of our presence, as the sea this
morning erased the footprints of the bird.

– Rachel Carson (1950)

IN THE INTRODUCTION'S EPIGRAPH, the poet Derek Walcott describes the ocean as 'a grey vault' that locks up monuments, battles, martyrs and memories, all things that are the stuff of history. His vision presents the ocean as an archive, storing not only historical evidence and commemorative markers but even history itself. Carson's invocation of waves obliterating footprints on sand, perhaps more familiar an image than Walcott's, might on the surface appear to refer to the timeless ocean. In equating time to waters sweeping away footprints, though, Carson suggests the intertwined history of people and oceans; waves would erase human footprints as readily as a bird's. Time to her resembles the ocean, but time subsumes both the sea and everything else, 'containing all that came before us, sooner or later sweeping us away on its flood.'[1] Walcott likewise posits a relationship between the ocean and time, but he asserts that, 'The sea is History.' How can this be?

This enigma is a subset of the larger puzzle of the interrelationships between people and the natural world. The environmental historian William Cronon used the example of Chicago and its relationship to its hinterlands to point out the futility of trying to untangle the lasting imprint of human actions from an unrecoverable, if not mythical, 'first nature' that existed before or apart from humans. Richard White's history of the Columbia River identifies it as an 'organic machine', a system modified by people and embedded with technologies that nevertheless continues to manifest its natural qualities. The ocean is equally 'second nature', strange as that claim may seem for a part of the planet so incredibly vast and forbidding. This book argues that today's ocean is a human ocean, and that the ocean has a history relevant to humanity starting when life evolved in the sea.

We may today be in the midst of a third discovery of the ocean, prompted by changes that have made the ocean and its depths more visible and culturally accessible. Mariners in the fifteenth and sixteenth centuries found connections between the seas that enabled the global circulation of commodities, people and ideas. This discovery set in motion the transformation of the ocean, achieved by the eighteenth century, from a space for human activity into an empty space that could be controlled by those with knowledge of it. Despite the continuing – indeed, expanding – economic importance of the ocean to the present, most people in the nineteenth and early twentieth centuries experienced a transformation of the sea from a place of work into a site for recreation or respite. This shift promoted the sense of the sea as separate from people and unaffected by human activities, rendering the ocean into a timeless place. The post-war frontier metaphor strengthened the perception of limitless resources and infinite possibility to apply science and technology to forge new uses for the ocean. When environmental activists first contemplated the ocean, its timelessness posed a fundamental contradiction that few acknowledged: how can a place outside history be involved with massive change and urgent problems? Only very recently has the ocean and its role in our world gripped the attention of mainstream media and ordinary people. That has begun to happen in conjunction with an emerging cultural shift in our understanding of the ocean, one that has begun to reverse the timelessness and separateness from humanity assigned to the sea in the past.

Space exploration, long viewed as a rival endeavour by ocean boosters, unexpectedly contributed a new appreciation of the ocean that, ironically, undermined the frontier metaphor that came to be so closely associated with both space and oceans. Although NASA's gaze was mostly not directed back towards the Earth, Apollo astronauts produced memorable and widely circulated images that called attention to the uniqueness of our planet. The 'Earthrise' shot taken in 1968, showing a vibrant Earth rising over the horizon of the dead Moon, had a dramatic effect on environmental thinking. This and other images of Earth from space emphasized its absolute boundedness and transformed the technological metaphor of 'Spaceship Earth' into an ecological one. The first photograph of the whole Earth, taken in 1972, struck viewers with how much more of the planet is ocean rather than land. Dubbed 'Blue Marble', this picture showed a vividly blue planet against a backdrop of black, lifeless space. As Arthur C. Clarke put it, 'How inappropriate to call this planet Earth, when it is quite clearly Ocean.'[2]

It took time for the new visibility of the ocean as a planetary feature to prompt reassessments of cultural ideas about the sea. In the grip of the terrestrial environmental movement, Spaceship Earth generated reflection on the finiteness of land-based resources but conveyed optimism about the prospect of human control through technology despite environmental concern. In the decades after the first Earth Day of 1970, the ocean gradually attracted the attention of a handful of activists and concerned scientists. In 1972 a small organization called the Delta Corporation was founded to protect the ocean, but its name did not include the word 'marine' until 1989, when it became the Center for Marine Conservation, and it became the Ocean Conservancy only in 2003. That year it was joined on the front lines of environmental activism by another ocean-focused organization, Blue Frontier Campaign. The dolphin-safe tuna movement gained traction only in the 1990s, as did efforts to manage the ballast water of cargo ships to prevent the introduction of non-native marine species to ports around the globe. Such expressions of concern were halting and usually focused on the particular issue at hand rather than paying environmental attention to the ocean itself.

View of the Earth photographed by the *Apollo 17* crew on 7 December 1972
as they travelled towards the Moon.

Two issues have finally disrupted the long-standing pattern of
episodic notice of parts of the ocean. The first in time was the shocking
closure in the 1990s of the centuries-old cod fishery in the northwest
Atlantic. As extraordinary as this closure was, it was followed by the
inconceivable failure of the stocks to recover as expected, or as experts
had confidently predicted, with fishing pressure gone. More than a
decade of study and reflection was needed to comprehend what had
happened. Overfishing of Atlantic cod had not simply reduced the availability of a valuable commercial resource; it actually altered the
ecosystem. Since then, numerous studies have confirmed that massive
fish population declines exist everywhere. In recent years, awareness
of fisheries-related issues has expanded beyond specialist audiences,
so that concerns about by-catch, damage to the sea floor by trawling

and fishing down the food chain appear in mainstream media and are discussed outside fishing and scientific communities. Jokes about jellyfish being served as a restaurant's 'catch of the day' reveal broadening cultural awareness and insight into complex ecological and economic oceanic realities.[3] The devastation to the cod fishery and the consequent massive-scale impact on the ocean caused by fishing finally eroded the tenaciously held view of the limitlessness of fish and the imperviousness of the ocean to human activity. Second has been the debate over global climate change. The science investigating this issue is, at present, revealing the extent to which the ocean drives forces globally. And the ocean and its coasts also feature prominently in the catalogue of expected planet-wide effects from climate change. More than 90 per cent of the excess heat from global warming is absorbed by the ocean, leading to warming of the water, shifts in the ranges of many species and the elimination of others. Ocean acidification, caused when the seas absorb excess carbon dioxide, has already begun altering the chemistry of seawater, notably reducing the amount of calcium carbonate that many

Marine biologist and photographer Hans Hillewaert's illustration, inspired by the discoveries of marine scientist Daniel Pauly, of fishing down the food web, using the North Sea ecosystem as an example.

marine organisms need to build their skeletons. Experts worry about drastic effects on shellfish such as oysters and clams, on shallow- and deep-sea corals, and even on tiny calcareous plankton, an important pillar at the base of the oceanic food web. The seriousness of these consequences motivates advocates of geo-engineering to propose controversial global-scale actions to mitigate them. Sprinkling iron or mineral dust into the sea to decrease or absorb carbon dioxide seems appealing as an easy fix, but such interventions are unproven and carry a significant risk of unintended deleterious consequences. Many scientists and environmentalists are unwilling to countenance global experiments, while technological optimists worry about not pursuing research in this area.

Sea-level rise, already apparent in many places, will affect certain coasts more than others. That, combined with the increased intensity of storms that develop over warmer ocean water, will require profound adaptation around the world, including the necessity for the citizens of low-lying island nations to relocate to new land in order to survive. Today most people on Earth live near coasts, occupying about 10 per cent of the Earth's terrestrial surface. Only in Africa do fewer people live along coasts or major rivers than in the interior, but even there migration from the countryside to coastal cities is changing that balance. Around the globe, sea-level rise will require increasingly crowded urban coastal communities to adapt even as some will be eliminated. Recreation on and along the water has expanded enormously, placing whale watching, pleasure boating, sport fishing, surfing, scuba diving and other such activities into direct competition with both traditional uses of coastal areas such as fisheries and new uses such as wind farms and liquid natural gas terminals.

The examples of overfishing and global climate change demonstrate that the ocean has changed, and is currently changing rapidly, in response to human actions. Some experts propose that the influence of human activities on Earth is significant enough to constitute a new geological period, the Anthropocene. Climate change caused by human activities affects ocean temperature, acidity and sea level. Overfishing and bottom trawling have profoundly altered marine ecosystems. Other uses of the ocean have similarly indelible effects. Deeper and more remote parts of the ocean have been more firmly drawn into the human realm. Long used as a dump site, the sea – including its surface, coasts and even the deep ocean floor – now harbours plastic, which does not biodegrade. Nylon ghost nets continue fishing indefinitely, marine animals ingest plastic bits mistaking them for food and chemicals from litter leach into seawater. The achievement of offshore oil drilling in ever greater depths prompts competition between users of different resources such as fish and oil. Accidents such as the 2010 Deepwater Horizon BP spill in the Gulf of Mexico carry some of the same shock value as earlier ocean disasters and raise awareness of economic and ecological damage. So long as prices remain high enough, ocean drilling continues and expands, although critics point to its contribution to global climate problems. And virtually no corner of the ocean is unaffected by noise produced by seismic technology used to prospect for oil, as well as by giant tanker and container ships and by active sonar for naval operations. Together such human-produced sound covers the ocean like an inescapable net, affecting marine mammals and fish who rely on sound for survival. Often imperceptible to human ears, this noise can disrupt feeding and other behaviours, and has been associated with whales and dolphins becoming stranded on a large scale.

While many of the futuristic uses imagined in the post-war period have not come to pass, the oceans remain important economically. Fish are the only major wild-caught food still widely consumed, providing critical protein to much of the world – and it is far from clear that aquaculture is going to solve the overfishing crisis. Other long-standing uses of marine life, including animal feed and fertilizer, have expanded. Bio-prospectors search for usable compounds from marine organisms such as algae, bacteria and invertebrates. Pirates once again threaten seafarers, endangering gargantuan cargo ships, yachts and even cruise ships. Warming at the poles is having conspicuous effects on

Plastic bags in the waters of Bunaken National Park, a marine park in North Sulawesi, Indonesia, imperil sea turtles that mistake them for food: one of the numerous dangers of plastic waste in the ocean.

high-latitude ecosystems, opening the once-mythical Northwest Passage and enabling commercial shipping as well as oil drilling in the Arctic, an area not yet governed under the international Law of the Sea regime. Although separate statistics are not kept to quantify ocean-related economic activities relative to land-based ones, numerous studies have demonstrated that industries and activities related to the sea form a significant part of economies from local to global levels.

Much, perhaps most, of the rhetoric and analysis of the environmental crisis of the ocean has been presented by scientists and policy experts, and there is no doubt that the very best possible science and technology are essential for understanding and addressing such problems. Yet the long delay in cultural recognition of ocean as needing and deserving protection similar to the land signals that more and better science, however important, is simply not enough. The changes in motion that have begun to illuminate the human relationship with the ocean will need to be extended.

One pivotal change is the erosion of the ocean's age-old opacity, as its vast extent and now-fathomable depths are becoming visible to any interested people with access to the Internet. Underwater filming of outstanding quality reveals marine creatures in their own habitats, even those in the deepest spots on Earth. Electronic tracking of individual sharks and sea turtles, in conjunction with Internet-based applications, allows people to follow them around the ocean. Citizen science opportunities, such as jellywatch.org, invite ocean visitors to report marine animal sightings for scientific use.[4] Animation of ocean currents etches these phenomena visibly so viewers encounter an image reminiscent of the early world maps whose oceans were crowded with textured surface water.[5] The technologies of buoys, remotely operated vehicles, satellites and autonomous underwater vehicles collect an almost incomprehensible amount of data about the ocean, dwarfing the data collected during all the world's oceanographic voyages combined. High in the air, travellers on planes with screens that show the landscape below now see oceans depicted with contours and named sea-floor features, whereas fliers before the closure of the Atlantic cod fisheries saw on the first generations of those screens only blank blue space in the oceans.

This increased visibility of the ocean enables a shift in who can be involved with it. Not just experts but ordinary people can recognize the sea as an essential part of our globe and our lives. We are familiar with thinking about the ocean in terms of its utility for human interests and needs: as a surface for transport or the extension of national power, a font for resources, an arena for war or a site for nostalgia and recreation. We may be on the verge of acknowledging and valuing the ocean on its own terms, of appreciating how much we depend on its mere existence. While ocean-related environmental issues are certainly attracting attention, they have done so in the past without gaining cultural traction. What has changed? Although the new-found visibility is instrumental, the third discovery of the ocean hinges on the recovery of its history.

Fisheries science, at first glance a seemingly unlikely source, has contributed a revolutionary and deeply historical insight derived from its attention to patterns of past catch statistics. In 1995 Daniel

Pauly pointed out that his cohort of researchers, and each preceding generation, tended to use the stock size and species composition of catches existing when they began their careers as the starting point for their research. He urged efforts to combat this 'shifting baseline syndrome' and to search for ways to lengthen our time frame. A reference point from the 1950s ignores fisheries depletions from the first half of the twentieth century. Even though information about the size of bluefin tuna caught in the 1920s or the number of Caribbean sea turtles harvested by a single vessel in the seventeenth century exists only anecdotally, the solution is not to ignore such facts about the oceans in the past, which is what has happened when scientists only use series of data collected during their working lifetimes. The shifting baselines concept has begun to reframe discussion of overfishing, and also of marine ecological restoration, to insist on the historicity of marine ecosystems, to expand the time range considered and to account for changes in how people have known the ocean over time.

Today's understanding of global climate dynamics similarly involves grappling with the deeply historical phenomenon of El Niño. The warm phase of a larger pattern of cyclical warm and cool sea surface temperatures in the equatorial Pacific, El Niño has near-global reach and its effects include droughts, mild winters or excessive rainfalls in different parts of the world. There is evidence that El Niño events have been occurring for thousands of years and that ancient civilizations, such as the Moche in pre-Columbian Peru, practised human sacrifice to try to alleviate the environmental devastation caused by their heavy rains. The name 'El Niño', meaning 'the boy child', reveals the awareness of Spanish colonists and coastal fishermen in the southeastern Pacific of warm currents that arrived around Christmas, brought heavy rains and appeared to drive fish away. Knowledge of this global oceanic phenomenon existed through the work and lives of people long ago. Particularly since the 1998 to 1999 event, perhaps the strongest on record, the pattern has attracted the attention of today's scientists who note that the last few events have been among the strongest of the past century.

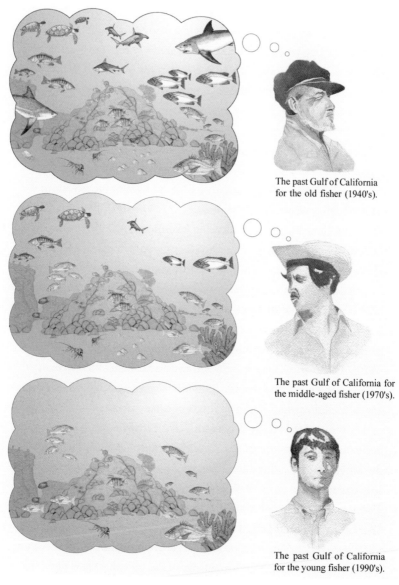

The past Gulf of California for the old fisher (1940's).

The past Gulf of California for the middle-aged fisher (1970's).

The past Gulf of California for the young fisher (1990's).

Shifting baselines mean that each generation views an already diminished marine ecosystem as pristine.

Rather suddenly, the ocean has come into focus not only among natural scientists discovering ocean acidification and trophic cascades, but among archaeologists, geographers, literary scholars and historians as well. Evidence of human prehistory and maritime history lies hidden on the ocean floor, and governments are moving to protect known sites, while underwater archaeologists are hopeful that submerged sites of ancient coastal habitation might be found and investigated. Geographers have explored differences between cultural perceptions of the ocean held by people around the world, and historians are newly conscious that the ocean environment and human activities connected to it demand, and deserve, their scrutiny.

Ethics joins history in forging an emerging new cultural perception of the ocean. The marine ecologist and activist Carl Safina calls for the embrace of a 'sea ethic' in parallel to the 'land ethic' principle articulated in Aldo Leopold's 1949 book of essays, *A Sand County Almanac*. This principle recognized the right of the natural world to exist regardless of its utility for humans. Leopold's moral call-to-arms had a profound effect on the formal environmental movement of the 1960s, although his terrestrial focus did nothing to distract the movement from its land orientation. Today's oceans include sanctuaries and protected areas as well as fisheries, reflecting our growing understanding that healthy marine ecosystems are more resilient in the face of climate challenges. Sanctuaries protect not only living creatures but now preserve submerged cultural heritage, another sign of the growing consciousness of the ocean as a historical site. In terms of geopolitical importance, the oceans are more vital now than they have ever been, a reality reflected in the increased tempo of discussions surrounding ocean policy, law of the sea and marine spatial planning.

Although the ocean has a natural history that far precedes us, the origin of life in the oceans, and especially our beginnings as a species, initiated the present narrative of ocean history, one that includes and involves us. People have always lived with and by the sea – not all people, of course – but all have been touched by trade, cultural beliefs and climate, each of which depends on the ocean. From the first discovery of the world ocean in the fifteenth century, the human connection with the sea has tightened through global trade, empire-building, industrialization and commodification of nature, by means of work, ambition, imagination and play. Knowledge of the sea has inspired and enabled new uses of its resources and waters even as it bolstered and expanded traditional maritime activities.

The humanities help to deal with the ocean's characteristic of tracklessness. Unlike the wagon wheels of settlers who traversed the Oregon Trail, which left ruts still visible in places across the western U.S. today, the wake of boats is ephemeral. Ordinary viewers cannot sense the past when they are on the ocean, although coastal events like storms and oil spills make visible the effects of some human activities on the sea. Whereas physical and chemical oceanographers can render the vast, trackless ocean comprehensible by measuring temperature differences, currents or minute amounts of particles, historians can use archival and other sources to recover activities that took place on and in oceans, activities which became invisible to the observer as soon as

A recreational sailboat in front of a cluster of wind turbines:
an example of the competing uses of ocean space in the 21st century.

the fishers' nets were hauled aboard, the ship's wake faded or the rising tide obliterated the footprints.

Darwin taught us in 1859 that we are part of nature. In some languages, and in the minds of many romantics and mariners, the sea and ships remain feminine, a cultural relic. Starting in 1941 with her first ocean book, *Under the Sea Wind*, Carson exercised great care to avoid anthropomorphism in her writing, an act that reminds us that the sea is neutral. *Lloyd's List*, the specialist newspaper providing shipping news since 1734, irked traditionalists when it changed its 278-year-old editorial practice in 2012. It replaced the feminine pronoun 'she' to refer to vessels with the neutral and business-like pronoun 'it'.[6] The culturally bound view of the sea and ships as feminine exists in tension with the apparently contradictory acceptance of ships as inanimate technologies subject to the forces of a dispassionate, natural ocean. This apparent conflict dissolves if we accept the sea as second nature, a conflation of natural forces and human constructs – that is, a human ocean.

If we are mindful of geological time, we recognize that we need the ocean more than it needs us. Indeed, it does not need us at all. Scientists predict that, if humans disappeared from Earth, reefs and most marine species would recover. Carson intended for her chronicle of the wonders and mysteries of interconnected physical and biological worlds to convey optimism for positive change, and many readers were, indeed, inspired by her message to active environmentalism. Yet humans in her story appeared as predators or destroyers. Following that cue, many environmental narratives lapse into tales of inevitable decline.

Until we recognize the ocean's past, and our inextricable relationship to it, we will not make much headway in changing that relationship for the better. The importance of the humanities, and the very power of metaphors such as 'frontier' or 'wilderness', hold out hope. We must jettison our perception of the ocean as a timeless place, apart from humans. We must transform our understanding of the sea to one bound with history and interconnected with humanity. Such a new vision, with new metaphors, can form the foundation for positive change.

REFERENCES

◞ONE A Long Sea Story

1 See, for example, 'A Hero's Reward', *The Mariners* (19 November 2008), www.themariners.org, and Robert J. Clark, 'uss Eugene A. Greene (DD-711/DDR-711) Reunion', The National Association of Destroyer Veterans, www.destroyers.org.

2 Letter from Charles Darwin to J. D. Hooker, 1 February [1871], Darwin Correspondence Project, www.darwinproject.ac.uk.

3 'Age of Oysters' is a coinage of J. Bret Bennington, a geologist at Hofstra University. See course notes, 'Geol 106, Mesozoic Marine Life' [n.d.], http://people.hofstra.edu.

◞TWO Imagined Oceans

1 J. O'Kane, trans., *The Ship of Sulaiman* (London, 1972), p. 163.

2 Barry Cunliffe, *Facing the Ocean: The Atlantic and its Peoples* (Oxford and New York, 2001), p. vii.

3 William R. Pinch, 'History, Devotion and the Search for Nabhadas of Galta', in *Invoking the Past: the Uses of History in South Asia*, ed. Daud Ali (Delhi and Oxford, 1999), pp. 367–99.

4 Epeli Hau'ofa, 'Our Sea of Islands,' in *A New Oceania: Rediscovering Our Sea of Islands*, ed. Eric Waddell, Vijay Naidu and Epeli Hau'ofa (Suva, Fiji, 1993), p. 11.

5 Thomas Gladwin, *East is a Big Bird: Navigation and Logic on Puluwat Atoll* (Cambridge, MA, 1970), pp. 39, 63; David Lewis, *We the Navigators: the Ancient Art of Landfinding in the Pacific*, 2nd edn (Honolulu, HI, 1997), pp. 297–311.

◞THREE Seas Connect

1 Walter Raleigh, *Judicious and Select Essayes and Observations by That Renowned and Learned Knight, Sir Walter Raleigh* (London, 1650), p. 20.

2 Ram P. Anand, *Origin and Development of the Law of the Sea* (The Hague, 1983), p. 83.
3 Jonathan Raban, ed., *The Oxford Book of the Sea* (Oxford and New York, 1993), p. 3.
4 William Bradford, *Of Plymouth Plantation, 1621–1647: The Complete Text*, ed. Samuel Eliot Morison (New York, 1952), p. 61.
5 William Strachey, Esq., 'A True Reportory of the Wracke, and Redemption of Sir Thomas Gates, Knight . . .', in *Hakluytus Posthumus, or Purchas his Pilgrimes . . .* by Samuel Purchas, vol. IV (London, 1625), p. 1735, as quoted in Jason W. Smith, 'The Boundless Sea', in 'Controlling the Great Common: Hydrography, the Marine Environment, and the Culture of Nautical Charts in the United States Navy, 1838–1903', PhD thesis, Temple University, 2012, p. 4.
6 Bradford, *Of Plymouth Plantation*, pp. 62 and 61.
7 George Gordon, Lord Byron, from 'Childe Harold's Pilgrimage' (1818).
8 Natasha Adamowsky, *The Mysterious Science of the Sea, 1775–1943* (London and New York, 2015), p. 24.
9 Louis Agassiz, from letter of 15 June 1849, quoted in Eugene Batchelder, *A Romance of the Sea Serpent: Or, The Icthyosaurus* (Cambridge, MA, 1849), p. 135.

↷ FOUR Fathoming All the Ocean

1 'Sea', in *Encyclopædia Britannica Vol. XIX* (Edinburgh, 1823), p. 64.
2 Advertisement in *Harper's Weekly* (16 October 1858), p. 671.
3 Field's watch fob is in the Judson Collection, Division of Political History, National Museum of American History, Smithsonian Institution.
4 Edmund Gosse, *Father and Son* (London, 1907), pp. 125–6.
5 W. R. Hughes, 'The Recent Marine Excursion Made by the Society to Teignmouth,' *Nature*, IX (29 January 1874), pp. 253–4.
6 Herman Melville, *White-Jacket* (New York, 1850), p. 105.
7 Letter from George Brown Goode to Spencer F. Baird, 1878, Smithsonian Institution Archives, Spencer Fullerton Baird Papers, RU 7002, Box 21.

↷ FIVE Industrial Ocean

1 Rudyard Kipling, *Captains Courageous* (New York, 1897), p. 40.
2 The term 'Gospel of Efficiency' comes from Samuel P. Hays, *Conservation and the Gospel of Efficiency: The Progressive Conservation*

Movement 1890–1920 (Cambridge, 1959), pp. 1–4, 27–48; Jennifer Hubbard, 'The Gospel of Efficiency and the Origins of MSY: Scientific and Social Influences on Johan Hjort and A. G. Huntsman's Contributions to Fisheries Science', in *A Century of Marine Science: The St. Andrews Biological Station*, ed. David Wildish Hubbard and Robert Stephenson (Toronto, 2016), pp. 78–117.

3 Frederic A. Lucas, 'Conservation of Whales', *New York Times* (1 November 1910), p. 3.

4 Remington Kellogg, 'Whales, Giants of the Sea', *National Geographic*, LXXVII/1 (January 1940), pp. 35–90, quote on p. 35.

5 Richard Henry Dana, Jr, *Two Years before the Mast*, ed. Thomas Philbrick (New York, 1981), pp. 161–2.

6 Henry David Thoreau, *Cape Cod* (Boston, MA, and New York, 1896), p. 85.

7 Emily Dickinson, in *The Oxford Book of the Sea*, ed. Jonathan Raban (Oxford and New York, 1993), pp. 256–7.

8 John R. Gillis, *The Human Shore: Seacoasts in History* (Oxford and New York, 1993), p. 128.

9 John Masefield, 'Sea Fever', in *The Collected Poems of John Masefield* (London, 1923), pp. 27–8.

10 Thoreau, *Cape Cod*, p. 85.

11 George Gordon, Lord Byron, from 'Childe Harold's Pilgrimage' (1818), in *Poems of Places: An Anthology in 31 Volumes*, ed. Henry Wadsworth Longfellow (Boston, MA, 1876–9); available at Bartleby.com, 2011, accessed 20 June 2017.

12 Michael Graham, *The Fish Gate* (London, 1943), p. 150.

13 Michael Graham, 'Harvests of the Sea', in *Man's Changing Role in the Face of the Earth*, ed. William L. Thomas Jr (Chicago, IL, 1956), p. 502.

14 Garrett Hardin, 'The Tragedy of the Commons', *Science*, CLXII (1968), pp. 1243–8.

15 Carmel Finley, *All the Fish in the Sea: Maximum Sustainable Yield and the Failure of Fisheries Management* (Chicago, IL, and London, 2011), pp. 88, 182.

SIX Ocean Frontier

1 Seabrook Hull, *The Bountiful Sea* (Englewood Cliffs, NJ, 1964), p. 221.

2 Advertisement by American Petroleum Institute, 'U.S. Oilmen Challenge the Sea', *Life*, XXXVI/23 (7 June 1954), p. 152.

3 Edwin L. Hamilton, 'The Last Geographic Frontier: The Sea Floor', *Scientific Monthly*, LXXXV/6 (December 1957), pp. 294–314.

4 Donald W. Cox, *Explorers of the Deep: Man's Future beneath the Sea* (Maplewood, NJ, 1968), p. 9.
5 Vannevar Bush, *Science the Endless Frontier: A Report to the President* (Washington, DC, 1945).
6 Alexander McKee, *Farming the Sea* (New York, 1969), p. 4.
7 Arthur C. Clarke, *The Challenge of the Sea* (New York, 1960), p. 111.
8 John L. Mero, *The Mineral Resources of the Sea* (Amsterdam, 1964); quote is in frontispiece caption.
9 John F. Kennedy, Letter to President of the Senate on Increasing the National Effort in Oceanography, 29 March 1961, Letter 100, The American Presidency Project, www.presidency.ucsb.edu.
10 Vernon Pizer, *The World Ocean: Man's Last Frontier* (Cleveland, OH, 1967), p. 137.
11 George Bond, 'New Development in High Pressure Living', *Archives of Environmental Health*, IX (1964), p. 311.
12 President's Science Advisory Committee, Panel on Oceanography, *Effective Use of the Sea* (Washington, DC, 1966), pp. 104–5.
13 Lockheed advertisement in 'Back Matter', *Science*, new ser. CLI/3715 (11 March 1966), p. 1311. Published by the American Association for the Advancement of Science (4 July 2009), available at www.jstor.org.
14 John Ludwigson, 'Law Comes to the Sea Floor', *Science News*, XCI/20 (20 May 1967), p. 474.
15 John E. Bardach, *Harvest of the Sea* (New York, 1968), p. 10.
16 Martin Ira Glassner, *Neptune's Domain: A Political Geography of the Sea* (Boston, MA, 1990), p. 5.
17 Roger Revelle, 'The Ocean', *Scientific American*, XXCCI/3 (September 1969), p. 56.

⌐ SEVEN Accessible Ocean

1 Arthur C. Clarke, *The Challenge of the Sea* (New York, 1960), p. 164.
2 Eric Hanauer, *Diving Pioneers: An Oral History of Diving in America* (San Diego, CA, 1994), p. 20.
3 Bill Barada, *Let's Go Diving: Illustrated Diving Manual* (Santa Ana, CA, 1965), pp. 4–5.
4 'A Carload of Fun Comes to Shelby, Ohio', *Life*, XLVII/26 (28 December 1959), pp. 93–5.
5 'Tucking in a Tadpole', *Life*, LI/2 (14 July 1961), p. 108.
6 Zale Parry, quoted in Hanauer, *Diving Pioneers*, p. 150.
7 Eugenie Clark, *Lady with a Spear* (New York, 1953), p. 209.

8 Rachel L. Carson, application to the Eugene F. Saxton Foundation, 1 May 1949, cited in Linda Lear, *Rachel Carson: Witness for Nature* (New York, 1997), p. 162.
9 Ibid., p. 203.
10 Barada, *Let's Go Diving*, p. 7.
11 Rachel Carson, *The Sea Around Us* (New York, 1961), p. 196.
12 Gary Kroll, *America's Ocean Wilderness: A Cultural History of Twentieth-Century Exploration* (Lawrence, KS, 2008).
13 Anon., 'Tap Pryor, Crusading Biologist, Frontiersman of the Sea', *Life*, LXIII/17 (27 October 1967), p. 45.
14 Blake Clark, 'Hawaii's Showcase of the Sea', *Reader's Digest*, XCIII (August 1968), p. 146.
15 American Cetacean Society, 'Gray Whale' factsheet, http://acsonline. org, accessed 17 July 2017.

Epilogue: Ocean as Archive, Sea as History

1 Rachel Carson, Field Notes, Nags Head, 9 October 1950, quoted in Linda Lear, *Rachel Carson: Witness for Nature* (New York, 1997), p. 185.
2 This quote is widely ascribed to Arthur C. Clarke but not found in any publication I could locate. The quote appears, ascribed to Clarke, in James E. Lovelock, 'Hands Up for the Gaia Hypothesis', *Nature*, CCCXLIV/6262 (8 March 1990), p. 102.
3 Randy Olson, 'No Seafood Grille 2' (13 June 2013), www.youtube.com.
4 A citizen science example: jellywatch.org.
5 'NASA Views Our Perpetual Ocean' (9 April 2012), www.nasa.gov.
6 Andrew Hibberd and Nicola Woolcock, 'Lloyd's List Sinks the Tradition of Calling Ships "She"' (21 March 2002), www.telegraph.co.uk.

BIBLIOGRAPHY

Adamowsky, Natascha, *The Mysterious Science of the Sea, 1775–1943*
    (London and New York, 2015)
Allen, David E., *The Naturalist in Britain* (Princeton, NJ, 1994)
Alpers, Edward A., *The Indian Ocean in World History* (Oxford, 2014)
Anand, R. P., *Origin and Development of the Law of the Sea* (The Hague, 1983)
Anderson, Katharine, and Helen M. Rozwadowski, eds, *Soundings and
    Crossings: Doing Science at Sea, 1880–1970* (Sagamore Beach, MA, 2016)
Arch, Jakobina K., *Bringing Whales Ashore: Oceans and the Environment of
    Early Modern Japan* (Seattle, 2018)
Armitage, David and Alison Bashford, eds, *Pacific Histories: Ocean, Land,
    People* (Houndmills, Basingstoke and New York, 2014)
Aziz, F., et al., 'Archaeological and Paleontological Research in Central
    Flores, East Indonesia: Results of Fieldwork 1997–98', *Antiquity*
    LXXIII/280 (15 June 1999), p. 273
Barber, Paul H., et al., 'Biogeography: A Marine Wallace Line?' *Nature*,
    CDVI/1679 (17 August 2000), pp. 692–3
Barthelmess, Klaus, 'Basque Whaling in Pictures, 16th–18th Century',
    Itsas Memoria. Revista de Estudios Marítimos del País Vasco, 6,
    Untzi Museoa-Museo Naval (Donostia/San Sebastián, 2009),
    pp. 643–67
Bascom, Willard, *A Hole in the Bottom of the Sea: The Story of the Mohole
    Project* (Garden City, NY, 1961)
Bedarnik, Robert G., 'The Earliest Evidence of Ocean Navigation',
    International Journal of Nautical Archaeology, XXVI (1997), pp. 183–91
—, 'Seafaring in the Pleistocene', *Cambridge Archaeological Journal*, XIII
    (2003), pp. 41–66
Behrman, Cynthia Fausler, *Victorian Myths of the Sea* (Athens, OH, 1977)
Berta, Annalisa, *Return to the Sea: The Life and Evolutionary Times of Marine
    Mammals* (Berkeley and Los Angeles, CA, 2012)
Bolster, W. Jeffrey, *The Moral Sea: Fishing the Atlantic in the Age of Sail*
    (Cambridge, MA, 2012)

Bond, George F., and Helen A. Siiteri, *Papa Topside: The Sealab Chronicles of Capt. George F. Bond, USN* (Annapolis, MD, 1993)

Bradley, Bruce, and Dennis Stanford, 'The North Atlantic Ice-Edge Corridor: A Possible Paleolithic Route to the New World', *World Archaeology*, XXXVI/4 (2004), pp. 459–78

Braun, David R., et al., 'Early Hominin Diet Included Diverse Terrestrial and Aquatic Animals 1.95 Ma in East Turkana, Kenya', *Proceedings of the National Academy of Sciences of the United States*, CVII/22 (1 June 2010), pp. 10002–7

Brinnin, John Malcolm, *Sway of the Grand Saloon: A Social History of the North Atlantic* (New York, 1971)

Brown, Chandros Michael, 'A Natural History of the Gloucester Sea Serpent: Knowledge, Power, and Culture of Science in Antebellum America', *American Quarterly*, XLII/3 (1990), pp. 402–36

Brown, P., et al., 'A New Small-bodied Hominin from the Late Pleistocene of Flores, Indonesia', *Nature*, CDXXXI/7012 (28 October 2004), pp. 1055–61

Brumm, Adam, et al., 'Hominins on Flores, Indonesia, by One Million Years Ago', *Nature*, CDLXIV/7289 (1 April 2010), pp. 748–52

Brunner, Bernd, *The Ocean at Home: An Illustrated History of the Aquarium* (London, 2011)

Burnett, D. Graham, 'Matthew Fontaine Maury's "Sea of Fire": Hydrography, Biogeography and Providence in the Tropics,' in *Tropical Visions in an Age of Empire*, ed. Felix Driver and Luciana Martins (Chicago, IL, 2005), pp. 113–34

——, *The Sounding of the Whale: Science and Cetaceans in the Twentieth Century* (Chicago, IL, and London, 2012)

Buschmann, Rainer F., *Oceans in World History* (Boston, MA 2007)

Campbell, Tony, 'Portolan Charts from the Late Thirteenth Century to 1500', in *Cartography in Medieval Europe and the Mediterranean, History of Cartography, Volume 1*, ed. J. B. Harley and David Woodward (Chicago, IL, 1987), pp. 371–463

Carlisle, Norman, *The Riches of the Sea* ([New York], 1967)

Carlson, Patricia Ann, ed., *Literature and Lore of the Sea* (Amsterdam, 1986)

Carlton, James T., 'Introduced Invertebrates of San Francisco Bay', in *San Francisco Bay: The Urbanized Estuary*, ed. T. John Conomos (San Francisco, CA, 1979), pp. 427–44

——, 'Transoceanic and Interoceanic Dispersal of Coastal Marine Organisms: The Biology of Ballast Water', *Oceanography and Marine Biology: Annual Review*, XXIII (1985), pp. 313–71

Carrier, Rick, and Barbara, *Dive* (New York, 1955), including Appendix B, 'The Diving Clubs', pp. 282–4 (repr. from *Skin Diver* magazine)

Chaplin, Joyce E., 'The Pacific before Empire, *c.* 1500–1800', in *Pacific Histories: Ocean, Land, People*, ed. David Armitage and Alison Bashford (Houndmills, Basingstoke and New York, 2014), pp. 53–74

Cheney, Cora, and Ben Partridge, *Underseas: The Challenge of the Deep Frontier* (New York, 1961)

Cipolla, Carlo, *Guns, Sails, and Empires: Technological Innovation and European Expansion, 1400–1700* (New York, 1965)

[Civic Education Service], *Underwater World* (Washington, DC, 1967)

Clarke, Arthur C., *The Challenge of the Sea* (New York, 1960)

——, *The Deep Range*, in *The Ghost from the Grand Banks and The Deep Range* (New York, 2001)

——, *Profiles of the Future: A Daring Look at Tomorrow's Fantastic World* (New York, 1967)

Coggins, Jack, *Hydrospace: Frontier beneath the Sea* (New York, 1966)

Cohen, Margaret, *The Novel and the Sea* (Princeton, NJ, and Oxford, 2010)

Conway, Erik M., 'Drowning in Data: Satellite Oceanography and Information Overload in the Earth Sciences', *Historical Studies in the Physical and Biological Sciences* XXXVII (2006), pp. 127–51

Cooper, A., and C. B. Stringer, 'Did the Denisovians Cross Wallace's Line?' *Science*, CCCXLII/6156 (18 October 2013), pp. 321–3

Corbin, Alain, *The Lure of the Sea: The Discovery of the Seaside in the Western World, 1750–1840*, trans. Jocelyn Phelps (Cambridge, 1994)

Corliss, J. B., et al., 'Submarine Thermal Springs on the Galápagos Rift', *Science*, CCIII/4385 (1979), pp. 1073–83

Cowan, Robert C., *Frontiers of the Sea* (New York, 1960)

Cox, Donald W., *Explorers of the Deep: Man's Future beneath the Sea* (Maplewood, NJ, 1968)

Cronon, William, *Nature's Metropolis: Chicago and the Great West* (New York, 1992)

Crosby, Alfred W., *The Columbian Exchange: The Biological and Cultural Consequences of 1492* (Westport, CT, 1972)

Cunliffe, Barry, *Europe between the Oceans: 900 BC to AD 1000* (New Haven, CT, 2011)

——, *Facing the Ocean: The Atlantic and its Peoples* (Oxford and New York, 2001)

Cunnane, Stephen C., *Survival of the Fattest: The Key to Human Brain Evolution* (Hackensack, NJ, 2005)

Cushman, Gregory T., *Guano and the Opening of the Pacific World: A Global Ecological History* (New York, 2013)

Davidson, Ian C., and Christina Simkanin, 'The Biology of Ballast Water 25 Years Later', *Biological Invasions*, XIV (2012), pp. 9–13

Dawson, Kevin, 'Enslaved Swimmers and Divers in the Atlantic World',
   Journal of American History, XCII/4 (2006), pp. 1327–55
Deacon, Margaret, *Scientists and the Sea, 1650–1900: A Study of Marine
   Science*, 2nd edn (Brookfield, VT, 1997)
Dingle, H., *Migration: The Biology of Life on the Move* (Oxford, 1996)
Dixon, E. J., 'Human Colonization of the Americas: Timing, Technology
   and Process', *Quaternary Science Reviews*, XX/1–3 (2001), pp. 277–99
Donovan, Arthur, and Joseph Bonney, *The Box that Changed the World:
   Fifty Years of Container Shipping: An Illustrated History* (East Windsor,
   NJ, 2006)
Dorsey, Kurkpatrick, *Whales and Nations: Environmental Diplomacy on the
   High Seas* (Seattle, WA, 2013)
Dreyer, Edward L., *Zheng He and the Oceans in the Early Ming Dynasty,
   1405–1433* (Boston, MA, 2006)
Dugan, James, *Man under the Sea* (New York, 1965)
Earle, Sylvia A., *Dive! My Adventures in the Deep Frontier* (Washington, DC,
   1999)
——, and Al Giddings, *Exploring the Deep Frontier: The Adventure of Man
   in the Sea* (Washington, DC, 1980)
Edmond, J. M., et al., 'Chemistry of Hot Springs on the East Pacific
   Rise and their Effluent Dispersal', *Nature*, CCVCVII/5863 (1982),
   pp. 187–91
Eilperin, Juliet, *Demon Fish: Travels through the Hidden World of Sharks*
   (New York, 2011)
Eldredge, Charles C., 'Wet Paint: Herman Melville, Elihu Vedder, and
   Artists Undersea', *American Art*, XI/2 (1997), pp. 106–35
Ellis, Richard, *Monsters of the Sea* (New York, 1995)
Erlandson, Jon M., 'The Archaeology of Aquatic Adaptations: Paradigms
   for a New Millennium', *Journal of Archaeological Research*, IX/4 (2001),
   pp. 287–350
——, et al., 'Paleoindian Seafaring, Maritime Technologies, and Coastal
   Foraging on California's Channel Islands', *Science*, CCCXXXI/6021
   (4 March 2011), pp. 1181–5, www.sciencemag.org
Everhart, Michael J., *Oceans of Kansas: A Natural History of the Western
   Interior Sea* (Bloomington, IN, 2005)
Fagan, Brian M., *Floods, Famines and Emperors: El Niño and the Fate of
   Civilizations* (New York, 1999)
Farris, William W., *Japan to 1600: A Social and Economic History*
   (Honolulu, HI, 2009)
Fernández-Armesto, Felipe, *Civilizations: Culture, Ambition, and the
   Transformation of Nature* (New York, 2001)

Finley, Carmel, *All the Fish in the Sea: Maximum Sustainable Yield and the Failure of Fisheries Management* (Chicago, IL, and London, 2011)

Finn, Bernard, *Submarine Telegraphy: The Grand Victorian Technology* (London, 1973)

Forbes, Edward, *The Natural History of the European Seas* Robert Godwin-Austen (London, 1859)

Fortey, Richard, *Trilobite: Eyewitness to Evolution* (New York, 2000)

Foulke, Robert, *The Sea Voyage Narrative* (New York, 1997)

Frank, Frederick S., and Diane Long Hoeveler, 'Introduction', *The Narrative of Arthur Gordon Pym of Nantucket* (Buffalo, NY, 2010), pp. 11–36

'Freedom of the Seas, 1609: Grotius and the Emergence of International Law' (exhibit marking the 400th anniversary of Hugo Grotius's *Mare Liberum*), 8 parts (22–23 October 2009), http://library.law.yale.edu

Gillis, John R., *The Human Shore: Seacoasts in History* (Chicago, IL, 2012)

——, *Islands of the Mind: How the Human Imagination Created the Atlantic World* (New York, 2003)

Gladwin, Thomas, *East is a Big Bird: Navigation and Logic on the Puluwat Atoll* (Cambridge, MA, 1995)

Glasscock, Carl B., *Then Came Oil: The Story of the Last Frontier* (Westport, CT, 1976, originally published in 1938)

Glassner, Martin Ira, *Neptune's Domain: A Political Geography of the Sea* (Boston, MA, 1990)

Gore, Rick, 'Who Were the Phoenicians?' *National Geographic* (October 2004), http://ngm.nationalgeographic.com

Gould, Stephen Jay, *Wonderful Life: The Burgess Shale and the Nature of History* (New York, 1990)

Gradwohl, Judith, and Michael L. Weber, *The Wealth of Oceans: Environment and Development on our Ocean Planet* (New York, 1995)

Grasso, Glenn, 'The Maritime Revival: Anti-modernism and the Maritime Revival, 1870–1940', PhD thesis, University of New Hampshire, 2009

Haag, Amanda Leigh, 'Patented Harpoon Pins Down Whale Age', *Nature* (19 June 2007), DOI:10.1038/news070618-6

Habu, Junko, 'Subsistence and Settlement', *Ancient Jomon of Japan* (Cambridge, 2004)

Hanauer, Eric, *Diving Pioneers: An Oral History of Diving in America* (San Diego, CA, 1994)

Hannigan, John A., *The Geopolitics of Deep Oceans* (Cambridge and Malden, MA, 2016)

Hau'ofa, Epeli, 'Our Sea of Islands', in *A New Oceania: Rediscovering Our Sea of Islands*, ed. Eric Waddell, Vijay Naidu and Epeli Hau'ofa (Suva, Fiji, 1993), pp. 2–16

Hellwarth, Ben, *Sealab: America's Forgotten Quest to Live and Work on the Ocean Floor* (New York, 2012)

Hellyer, David, 'Goggle Fishing in Californian Waters', *National Geographic*, XCV (May 1949), pp. 615–32

Helvarg, David, *Blue Frontier: Saving America's Living Seas* (New York, 2001)

Hinrichsen, Don, *Coastal Waters of the World: Trends, Threats and Strategies* (Washington, DC, 1998)

Höhler, Sabine, *Spaceship Earth in the Environmental Age, 1960–1990* (New York and London, 2015)

Holm, Poul, 'World War II and the "Great Acceleration" of North Atlantic Fisheries', *Global Environment*, X (2012), pp. 66–91

Howarth, David, *Dhows* (New York, 1977)

Hoyt, E., 'Whale Watching', in *Encyclopedia of Marine Mammals*, 2nd edn, ed. W. F. Perrin, B. Würsig and J.G.M. Thewissen (San Diego, CA, 2009), pp. 1219–23

Hubbard, Jennifer, 'Changing Regimes: Governments, Scientists and Fishermen and the Construction of Fisheries Policies in the North Atlantic, 1850–2010', pp. 129–68, in *A History of North Atlantic Fisheries, vol. 2, From the 1850s to the Early Twenty-First Century*, ed. David J. Starkey and Ingo Heidbrink (Bremen, 2012)

——, 'The Gospel of Efficiency and the Origins of MSY: Scientific and Social Influences on Johan Hjort and A. G. Huntsman's Contributions to Fisheries Science', in *A Century of Marine Science: The St Andrews Biological Station*, ed. Jennifer Hubbard, David Wildish and Robert Stephenson (Toronto, 2016), pp. 78–117

Huler, Scott, *Defining the Wind: The Beaufort Scale, and How a 19th-century Admiral Turned Science into Poetry* (New York, 2004)

Hull, Seabrook, *The Bountiful Sea* (Englewood Cliffs, NJ, 1964)

Hunt, Terry L., 'Rethinking the Fall of Easter Island', *American Scientist*, XCIV/5 (2006), pp. 412–19

Igler, David, *The Great Ocean: Pacific Worlds from Captain Cook to the Gold Rush* (Oxford and New York, 2013)

Irwin, Geoffrey, *The Prehistoric Exploration and Colonisation of the Pacific* (Cambridge, 1992)

Jackson, Jeremy B. C., et al., 'Historical Overfishing and the Recent Collapse of Coastal Ecosystems', *Science*, CCXCIII (27 July 2001), pp. 629–38

——, with Karen E. Alexander and Enric Sala, *Shifting Baselines: The Past and Future of Ocean Fisheries* (Washington, DC, 2011)

Jennings, Christian, 'Unexploited Assets: Imperial Imagination, Practical Limitations and Marine Fisheries Research in East Africa, 1917–1953', in *Science and Empire: Knowledge and Networks of Science across the*

British Empire, 1800–1970, ed. B. Bennett and J. Hodge (Houndmills, Basingstoke and New York, 2011), pp. 253–74

Jha, Alok, *The Water Book* (London, 2015)

Johannes, R. E., *Words of the Lagoon: Fishing and Maritime Lore in the Palau District of Micronesia* (Berkeley, CA, 1981)

Kani, Hiroaki, 'Fishing with Cormorant in Japan', pp. 569–83, in *The Fishing Culture of the World*, vol. I, ed. Béla Gunda (Budapest, 1984)

Kazar, John Dryden, 'The United States Navy and Scientific Exploration', PhD thesis, University of Massachusetts, 1973

Komatsu, Masayuki, and Shigeko Misaki, *Whales and the Japanese: How We Have Come to Live in Harmony with the Bounty of the Sea* (Tokyo, 2003); see especially, 'Whaling in Japan – from the Jomon to the Edo Period', pp. 45–54

Kroll, Gary, *America's Ocean Wilderness: A Cultural History of Twentieth-century Exploration* (Lawrence, KS, 2008)

Kurlansky, Mark, *The Basque History of the World: The Story of a Nation* (New York, 2001)

——, *Cod: A Biography of the Fish that Changed the World* (New York, 1997)

Kylstra, Peter H., and Arend Meerburg, 'Jules Verne, Maury and the Ocean', *Proceedings of the Royal Society of Edinburgh*, (B) LXXII/25 (1972), pp. 243–51

Labaree, Benjamin W., 'The Atlantic Paradox', in *The Atlantic World of Robert G. Albion*, ed. Labaree (Middletown, CT, 1975), pp. 195–217

Leamer, Robert B., Wilfred H. Shaw and Charles F. Ulrich, *Bottoms Up Cookery* (Gardena, CA, 1971)

Lear, Linda, *Rachel Carson: Witness for Nature* (New York, 1997)

Levathes, Louise, *When China Ruled the Seas: The Treasure Fleet of the Dragon Throne, 1405–1433* (Oxford, 1994)

Lewis, David, *We the Navigators: The Ancient Art of Landfinding in the Pacific*, 2nd edn, ed. Sir Derek Oulton (Honolulu, HI, 1994)

Lewis, Martin W., 'Dividing the Ocean Sea', *Geographical Review*, LXXXIX/2 (April 1999), pp. 188–214

——, and Kären Wigen, *The Myth of Continents: A Critique of Metageography* (Berkeley, CA, 1997)

Lieberman, Daniel E., 'Further Fossil Finds from Flores', *Nature*, CDXXXVII/7061 (13 October 2005), pp. 957–8

Little, Crispin T. S., 'The Prolific Afterlife of Whales', *Scientific American*, CCCII/2 (February 2010), pp. 78–84

Long, Edward John, *New Worlds of Oceanography* (New York, 1965)

McCurry, Justin, 'Ancient Art of Pearl Diving Breathes its Last' (24 August 2006), www.theguardian.com

Mackay, David, *In the Wake of Cook: Exploration, Science and Empire, 1780–1801* (New York, 1985)

McKenzie, Matthew G., *Clearing the Coastline: The Nineteenth-century Ecological and Cultural Transformation of Cape Cod* (Hanover and London, 2010)

McLean, Donald A., *The Sea: A New Frontier* (Pasadena, CA, 1967)

MacLeod, Roy, and Philip F. Rehbock, eds, *Nature in Its Greatest Extent: Western Science in the Pacific* (Honolulu, HI, 1988)

Mahadevan, Kumar, 'Mote Marine Laboratory: Exploring the Secrets of the Sea since 1955' (19 November 2010), http://mote.org

Malinowski, Bronislaw, *The Argonauts of the Western Pacific* (London, 1922)

Marean, Curtis W., et al., 'Early Human Use of Marine Resources and Pigment in South Africa during the Middle Pleistocene', *Nature*, CDXLIX/7164 (18 October 2007), pp. 905–8

Marx, Robert F., *The History of Underwater Exploration* (New York, 1978)

Marx, Wesley, *The Frail Ocean* (New York, 1967)

Matsen, Brad, *Jacques Cousteau: The Sea King* (New York, 2009)

Menard, William H., ed., *Oceans, Our Continuing Frontier* (Berkeley, CA, 1976)

Mentz, Steve, *At the Bottom of Shakespeare's Ocean* (London, 2009)

Mero, John L., *The Mineral Resources of the Sea* (Amsterdam, 1964)

Milam, Erika Lorraine, 'Dunking the Tarzanists: Elaine Morgan and the Aquatic Ape Theory', in *Outsider Scientists: Routes to Innovation in Biology*, ed. Oren Harman and Michael R. Dietrich (Chicago, IL, 2013), pp. 223–47

Miller, James W., and Ian G. Koblick, *Living and Working in the Sea*, 2nd edn (Plymouth, VT, 1995)

Miller, Sam, 'Skin Diver Magazine', Legends of Diving Articles (2006), www.internationallegendsofdiving.com

Mills, Eric L., 'Edward Forbes, John Gwyn Jeffreys, and British Dredging before the "Challenger" Expedition', *Journal of the Society for the Bibliography of Natural History*, VIII/4 (1978), pp. 507–36

Mitman, Gregg, *Reel Nature: America's Romance with Wildlife on Film* (Seattle, WA, 2009)

Mojzsis, S. J., et al., 'Evidence for Life on Earth by 3800 Million Years Ago', *Nature*, CDXLIX/6604 (1996), pp. 55–9

Mollat du Jourdin, Michele, *Europe and the Sea*, trans. Teresa Lavendar Fagan (Oxford, 1993)

Morwood, M. J., et al., 'Further Evidence for Small-bodied Hominins from the Late Pleistocene of Flores, Indonesia', *Nature*, CDXXXVII/7061 (13 October 2005), pp. 1012–17

Moseley, Michael Edward, *The Maritime Foundations of Andean Civilization* (Menlo Park, MA, 1975)

Mowat, Farley, *Sea of Slaughter* (Boston, MA, and New York, 1984)

Nielsen, Julius, et al., 'Eyelens Radiocarbon Reveals Centuries of Longevity in the Greenland Shark (*Somniosus microcephalus*)', *Science*, CCCLIII/6300 (12 August 2016), pp. 702–4

Norton, Trevor, *Stars beneath the Sea: The Pioneers of Diving* (New York, 1999)

Nukada, Minoru, 'Historical Development of the Ama's Diving Activities', in *Physiology of Breath-hold Diving and the Ama of Japan; Papers Presented at a Symposium August 31 to September 1, 1965*, ed. Hermann Rahn and Tetsuro Yokoyama (Washington, DC, 1965), pp. 27–40

O'Dell, Scott, *Island of the Blue Dolphins* (New York, 1960)

Paine, Lincoln, *The Sea and Civilization: A Maritime History of the World* (New York, 2013)

Parrott, Daniel S., *Tall Ships Down: The Last Voyages of the Pamir, Albatross, Marques, Pride of Baltimore, and Maria Asumpta* (Camden, ME, 2003)

Parry, J. H., *The Discovery of the Sea* (Berkeley, CA, 1981)

Pastore, Christopher L., *Between Land and Sea: The Atlantic Coast and the Transformation of New England* (Cambridge, MA, 2014)

Pauly, Daniel, 'Anecdotes and the Shifting Baseline Syndrome of Fisheries', *Trends in Ecology and Evolution*, X (10 October 1995), p. 430

Payne, Brian J., *Fishing a Borderless Sea: Environmental Territorialism in the North Atlantic, 1818–1910* (East Lansing, MI, 2010)

Pearson, Michael, *The Indian Ocean* (London and New York, 2003)

Perkins, Sid, 'As the Worms Churn', *Science News* (23 October 2009), www.sciencenews.org

Philbrick, Thomas, *James Fenimore Cooper and the Development of American Sea Fiction* (Cambridge, MA, 1961)

Pinch, William R., 'History, Devotion, and the Search for Nabhadas of Galta', in *Invoking the Past: The Uses of History in South Asia*, ed. Daud Ali (Delhi and Oxford, 1999), pp. 367–99

Pinchot, Gifford B., 'Whale Culture – A Proposal', *Perspectives in Biology and Medicine*, X/1 (1966), pp. 33–43

Pomeroy, Robert S., Nataliya Plesha and Umi Muawanah, 'Valuing the Coast: Economic Impact of Connecticut's Maritime Industry', *Connecticut Sea Grant Report* (March 2013), http://seagrant.uconn.edu

Poole, Robert, *Earthrise: How Man First Saw the Earth* (New Haven, CT, 2010)

Poulsen, Bo, *Global Marine Science and Carlsberg: The Golden Connection of Johannes Schmidt (1877–1933)* (Leiden, 2016)

Pratt, Joseph A., Tyler Priest and Christopher Castaneda, 'Inner Space Pioneer: Taylor Diving and Salvage', in *Offshore Pioneers: Brown & Root and the History of Offshore Oil and Gas* (Houston, TX, 1997), pp. 137–57

Preston, Diana and Michael, *A Pirate of Exquisite Mind: Explorer, Naturalist and Buccaneer: The Life of William Dampier* (New York, 2004)

Priest, Tyler, *The Offshore Imperative: Shell Oil's Search for Petroleum in Postwar America* (College Station, TX, 2007)

Pringle, Heather, 'Traces of Ancient Mariners Found in Peru', *Science*, CCLXXXI/5384 (1998), pp. 1775–7

Pryor, Karen, *Lads Before the Wind: Adventures in Porpoise Training* (New York, 1975)

Raban, Jonathan, ed., *The Oxford Book of the Sea* (Oxford and New York, 1993)

Rehbock, Philip F., 'The Early Dredgers: "Naturalizing" in British Seas, 1830–1850', *Journal of the History of Biology*, XII (1979), pp. 293–368

——, ed., *At Sea with the Scientifics: The 'Challenger' Letters of Joseph Matkin* (Honolulu, HI, 1992)

——, 'Huxley, Haeckel, and the Oceanographers: The Case of *Bathybius haeckelii*', *Isis*, LXVI (1975), pp. 504–33

Reid, Joshua, 'Marine Tenure of the Makahs', in *Indigenous Knowledge and the Environment in Africa and North America*, ed. David M. Gordon and Shephard Krech III (Athens, OH, 2012), pp. 243–58

Reidy, Michael S., 'From the Oceans to the Mountains: Spatial Science in an Age of Empire', in *Knowing Global Environments: New Historical Perspectives on the Field Sciences*, ed. Jeremy Vetter (New Brunswick, NJ, 2010), pp. 17–38

——, *Tides of History: Ocean Science and Her Majesty's Navy* (Chicago, IL, 2008)

——, with Gary Kroll, and Erik M. Conway, *Exploration and Science: Social Impact and Interaction* (Santa Barbara, CA, 2007)

——, and Helen M. Rozwadowski, 'The Spaces in Between: Science, Ocean, Empire', *Isis*, CV/2 (2014), pp. 338–51

Rice, A. L., 'The Oceanography of John Ross' Arctic Expedition of 1818: A Reappraisal', *Journal of the Society for the Bibliography of Natural History*, VII/3 (1975), pp. 291–319

——, and J. B. Wilson, 'The British Association Dredging Committee: A Brief History', in *Oceanography: The Past*, ed. Mary Sears and Daniel Merriman (New York, 1980), pp. 373–85

Richardson, Philip L., 'The Benjamin Franklin and Timothy Folger Charts of the Gulf Stream', in *Oceanography: The Past*, ed. Mary Sears and Daniel Merriman (New York, 1980), pp. 703–17

Rick, Torbin and Jon Erlandson, *Human Impacts on Ancient Marine Ecosystems: A Global Perspective* (Berkeley, CA, 2008)

Rieser, Alison, *The Case of the Green Turtle: The Uncensored History of a Conservation Icon* (Baltimore, MD, 2012)

Ritvo, Harriet, *The Playtypus and the Mermaid and Other Figments of the Classifying Imagination* (Cambridge, MA, 1997)

Roark, E. Brendan, et al., 'Extreme Longevity in Proteinaceous Deep-sea Corals', *Proceedings of the National Academy of Sciences of the United States of America*, CVI/13 (30 March 2009), pp. 5204–8

Roberts, Callum, *Ocean of Life: The Fate of Man and the Sea* (New York, 2013)

—, *The Unnatural History of the Sea* (Washington, DC, 2008)

Robinson, Michael F., 'Reconsidering the Theory of the Open Polar Sea', in *Extremes: Oceanography's Adventures at the Poles*, ed. Keith R. Benson and Helen M. Rozwadowski (Sagamore Beach, MA, 2007)

Robinson, Samuel A., *Ocean Science and the British Cold War State* (London, 2018)

Roney, J. Matthew, 'Taking Stock: World Fish Catch Falls to 90 Million Tons in 2012', Earth Policy Institute (19 November 2012), www.earth-policy.org

Rossby, H. Thomas, and Peter Miller, 'Ocean Eddies in the 1539 Carta Marina by Olaus Magnus', *Oceanography*, XVI/2 (2003), pp. 77–88

Rozwadowski, Helen M., 'Arthur C. Clarke and the Limitations of the Ocean as a Frontier', *Environmental History*, XVII/3 (2012), pp. 578–602

—, 'Engineering, Imagination, and Industry: Scripps Island and Dreams for Ocean Science in the 1960s', in *The Machine in Neptune's Garden*, ed. Rozwadowski and David van Keuren (Canton, MA, 2004), pp. 325–52

—, *Fathoming the Ocean: The Discovery and Exploration of the Deep Sea* (Cambridge, MA, 2005)

—, 'From Danger Zone to World of Wonder: The 1950s Transformation of the Ocean's Depths', *Coriolis: Interdisciplinary Journal of Maritime Studies*, IV/1 (2013), pp. 1–20

—, 'Oceans: Fusing the History of Science and Technology with Environmental History', in *A Companion to American Environmental History*, ed. Douglas Cazaux Sackman (Malden, MA, 2010), pp. 442–61

—, 'Playing by – and on and under – the Sea: The Importance of Play for Knowing the Ocean', in Jeremy Vetter, ed., *Knowing Global Environments: New Historical Perspectives on the Field Sciences* (New Brunswick, NJ, 2010), pp. 162–89

—, 'Scientists Writing and Knowing the Ocean', in *The Sea and Nineteenth-Century Anglophone Literary Culture*, ed. Steve Mentz and Martha Elena Rojas (London, 2017)

—, *The Sea Knows No Boundaries: A Century of Marine Science under ICES* (Seattle, WA, and London, 2002)

Rudwick, Martin J. S., *Scenes from Deep Time: Early Pictorial Representations of the Prehistoric World* (Chicago, IL, 1992)

Safina, Carl, *Song for the Blue Ocean: Encounters along the World's Coasts and beneath the Seas* (New York, 1998)

Sahrhage, Dietrich and Johannes Lundbeck, *A History of Fishing* (Berlin, 1992)

Samuel, Lawrence R., *The End of Innocence: The 1964–1965 New York World's Fair* (Syracuse, NY, 2007)

Schopf, William J., 'Microfossils of the Early Archaean Apex Chert: New Evidence of the Antiquity of Life', *Science*, CCLX/5108 (30 April 1993), pp. 640–46

—, ed., *Life's Origin: The Beginnings of Biological Evolution* (Berkeley and Los Angeles, CA, 2002)

Seelye, John, 'Introduction', *Arthur Gordon Pym, Benito Cereno, and Related Writings* (New York, 1967)

Sheffield, Suzanne Le-May, *Revealing New Worlds: Three Victorian Women Naturalists* (London and New York, 2001), pp. 13–74

Shimamura, Natsu, 'Abalone', *The Tokyo Foundation* (12 May 2009), www.tokyofoundation.org

Smith, Andrew F., *American Tuna: The Rise and Fall of an Improbable Food* (Berkeley, CA, 2012)

Smith, Jason W., *To Master the Boundless Sea: The U.S. Navy, the Marine Environment and the Cartography of Empire* (Chapel Hill, NC, 2018)

Smithsonian Institution, National Museum of Natural History, Ocean Portal Team, 'Deep-sea Corals' [n.d.], http://ocean.si.edu, accessed 15 May 2017

Sobel, Dava, *Longitude: The True Story of the Lone Genius Who Solved the Greatest Scientific Problem of His Time* (New York, 1995)

Soini, Wayne, *Gloucester's Sea Serpent* (Charleston, NC, and London, 2010)

Soule, Gardner, *Undersea Frontiers* (New York, 1968)

Spilhaus, Athelstan, *Turn to the Sea* (Racine, WI, 1962)

Sponsel, Alistair, *Darwin's Evolving Identity: Adventure, Ambition and the Sin of Speculation* (Chicago, 2018)

Steele, Teresa E., 'A Unique Hominin Menu Dated to 1.95 Million Years Ago', *Proceedings of the National Academy of Sciences of the United States of America*, CVII/24 (15 June 2010), pp. 10771–2, www.pnas.org

Stein, Doug, '"Whale of a Tale": George H. Newton and the Cruise of the Inland Whaling Association', *The Log of Mystic Seaport*, XL/2 (1988), pp. 39–49

Steinberg, Philip E., *The Social Construction of the Ocean* (Cambridge, 2001)

Sténuit, Robert, *The Deepest Days*, trans. Morris Kemp (New York, 1966)

Stopford, Martin, 'How Shipping Has Changed the World and the Social Impact of Shipping', paper delivered to the Global Maritime Environmental Congress, SSM Hamburg, 7 September 2010, http://ec.europa.eu

Stow, Dorrik, *Vanished Ocean: How Tethys Reshaped the World* (Oxford, 2012)

Stringer, C. B. et al., 'Neanderthal Exploitation of Marine Mammals', *Proceedings of the National Academy of Sciences of the United States*, cv/38 (23 September 2008), pp. 14319–24

Sweeney, John, *Skin Diving and Exploring Underwater* (New York, 1955)

Tesch, Frederich W., *The Eel*, 5th edn (Oxford, 2003)

Toussaint, Auguste, *History of the Indian Ocean*, trans. J. Guicharnaud (Chicago, IL, 1966)

Turner, Frederick Jackson, *The Frontier in American History* (New York, 1996)

Valle, Gustav Dalia, et al., *Skin and Scuba Diving*, Athletic Institute Series (New York, 1965)

Van Duzer, Chet, *Sea Monsters on Medieval and Renaissance Maps* (London, 2013)

Van Keuren, David K., 'Breaking New Ground: The Origins of Scientific Ocean Drilling', in *The Machine in Neptune's Garden: Historical Perspectives on Technology and the Marine Environment*, ed. David K. Van Keuren and Helen M. Rozwadowski (Canton, MA, 2004), pp. 183–210

Verlinden, Charles, 'The Indian Ocean: The Ancient Period and the Middle Ages', in Satish Chandra, *The Indian Ocean: Explorations in History, Commerce and Politics* (New Delhi and London, 1987)

Vetter, Richard C., *Oceanography: The Last Frontier* (New York, 1973)

Vickers, Daniel, *Farmers and Fishermen: Two Centuries of Work in Essex County, Massachusetts, 1630–1850* (Chapel Hill, NC, 1994)

Viviano, Frank, 'China's Great Armada, Admiral Zheng He', *National Geographic* (July 2005), http://ngm.nationalgeographic.com

Waldron, Arthur, *The Great Wall of China: From History to Myth* (Cambridge, 1992)

Walker, Timothy, 'European Ambitions and Early Contacts: Diverse Styles of Colonization, 1492–1700', in *Converging Worlds: Communities and Cultures in Colonial America*, ed. Louisa A. Breen (New York, 2012), pp. 16–51

Wendt, Henry, *Envisioning the World: The First Printed Maps, 1472–1700* (Santa Rosa, CA, 2010)

Westwick, Peter, and Peter Neushul, *The World in the Curl: An Unconventional History of Surfing* (New York, 2013)

White, Richard, *Organic Machine: The Remaking of the Columbia River* (New York, 1996)

Whitfield, Peter, *The Charting of the Oceans: Ten Centuries of Maritime Maps* (Portland, OR, 1996)

Wilford, John Noble, 'Key Human Traits Tied to Shellfish Remains', *The New York Times* (18 October 2007), www.nytimes.com

Witt-Miller, Harriet, 'The Soft, Warm, Wet Technology of Native Oceania', *Whole Earth*, LXXII (1991), pp. 64–9

Witzel, Michael, 'Water in Mythology', *Daedalus*, CXLIV/3 (2015), pp. 18–26

Woodward, David, and G. Malcolm Lewis, eds, *The History of Cartography, Volume 2, Book 3: Cartography in the Traditional African, American, Arctic, Australian, and Pacific Societies* (Chicago, IL, 1998)

Worm, Boris, Heike K. Lotze and Ransom A. Myers, 'Predator Diversity Hotspots in the Blue Ocean', *Proceedings of the National Academy of Sciences of the United States of America*, C/17 (19 August 2003), pp. 9884–8

Zelko, Frank S., *Make It a Green Peace: The Rise of Countercultural Environmentalism* (New York and Oxford, 2013)

Zincavage, David, 'Mapping Doggerland', 'Never Yet Melted' (11 July 2008), neveryetmelted.com

ACKNOWLEDGEMENTS

This book emerged from my research and teaching over more than a decade as part of the University of Connecticut's Maritime Studies Program. My faculty colleagues have challenged me to expand my interests and questions into many disciplines and lent their expertise generously. A fellowship at UConn's Humanities Institute set me off on the journey that resulted in this book. My students have convinced me of the heightened importance of the ocean in our world. Special thanks go to Russ Lycan, for his research on submerged sites archaeology, and Nathan Adams, for his research on menhaden, as well as to students in several History of the Ocean courses who experienced my initial efforts to put my argument and the story I relate here into its present form.

A number of colleagues at UConn and elsewhere spoke with me either about their own work as I began this study, or about mine in its very early stages, and have influenced this project in ways they may not know. I thank Katey Anderson, Matthiew Burnside, Peter Drakos, Carmel Finley, Ann Downer Hazel, Christian Jennings, Joy McCann, Michael Reidy, David Robinson, Josh Smith and Nancy Quam-Wickham. I am also happily indebted to many colleagues at The Mystic Seaport Museum with whom I have worked to offer public history courses, collaborate on research and teaching, and develop cooperative programming which animates the formal relationship (a Memorandum of Understanding) between the Museum and UConn. I have benefited especially from Elysa Engelman's insight and excellence as a historian. I also wish to thank Maia Sacca-Schaeffer, a Williams College student who provided research assistance in summer 2014.

I have the incredible good fortune to be part of a writing group that meets regularly at UConn's Avery Point campus and occasionally at the Firehouse. Pam Bedore, Susan Lyons and Anita Duneer are the regulars, but we enjoy the occasional company of others. I thank them for the hours of quiet concentration as well as the wide-ranging discussions about the writing process, our various projects and many other subjects as well. The chocolate and cookies are, of course, also inspiring. I have, in addition, benefited from the university-wide writing retreats supported by various offices at UConn that take place several times a semester at the Avery Point campus. Such institutional support for writing affirms, amid the day-to-day press of students who need attention and urgent university business, that scholarship remains a core part of faculty work. I am also grateful for support from UConn's College of Liberal Arts and Sciences Book Fund, which paid for some of the illustrations, and I thank Jorge Torre Cosio for his extraordinary efforts in finding the shifting baseline image.

The opportunity to share ideas with colleagues far and near is the greatest privilege of scholarship. My ideas about writing the ocean into history have developed over more than a decade, but I've presented parts of this book formally at a number of institutions where I've been invited to visit and lecture. The audiences have challenged me, sharpened my arguments and guided me in areas outside my expertise. I thank the European Society for Environmental History (Turku, 2011); organizers of the 2012 Blue Marble outreach event in San Diego, co-sponsored by the History of Science Society and Scripps Institution of Oceanography; members of the Mellon Sawyer Seminar at University of California, Santa Barbara (2014); participants of the Cornell Contested Global Landscapes Project (2014); scientists and others who attended the Maritime Research Symposium organized to celebrate the Royal Swedish Academy of Sciences 275th Jubilee; colleagues I met at the 2015 Underwater Realm Workshop at Stanford University; faculty and administrators involved with the Memorial University Arts on Oceans lecture in 2015; colleagues I met at the Royal Institute of Technology in Sweden in 2016; participants at the 2016 Aquatic Histories conference at the University of Tübingen's Institute for Eastern European History and Area Studies; scholars at the Rachel Carson Center in Munich, where I visited in 2016; and participants of the Deep Time, Deep Waters workshop, Institute for Advanced Studies in the Humanities, University of Edinburgh, in 2017.

Three chapters and sections of others are based mainly on my own research, but, given the span of time and geography covered, I have obviously relied on the scholarship of many others to craft this history. A number of colleagues have generously read portions of the manuscript, helping to identify mistakes and weaknesses and generally strengthening the book. Any remaining faults are certainly not theirs, but mine. I especially thank John Gillis, whose work often intersected in interesting ways with mine even before we met and who has offered much encouragement and assistance. Other readers whose help and ideas I wish to acknowledge include Jacobina Arch, Peter Auster, Mary K. Bercaw Edwards, Kurk Dorsey, Anita Duneer, Jon Erlandson, Marta Hanson, Penelope Hardy, Jennifer Hubbard, Stephen Jones, Brendan Kane, Adam Keul, Susan Lyons, Vijay (William) Pinch, Michael Robinson, Nancy Shoemaker, Timothy Walker and Daniel Zizzamia. I owe you all a beer, or something else of your choice, and I look forward to many happy hours discharging this debt.

Last, but really first and always, I thank my entire family. My parents raised me and my siblings on the shores of Lake Erie, where I learned to sail small boats, read sea stories and imagine the ocean. I never lived near the ocean until I moved to Connecticut in 2003. My children, Thad and Meg, who have grown up on the coast, and also my niece Hannah and nephews Landon and Jackson, remind me that my most important readers are the generations that will shape our human relationship with the ocean into the future. I am grateful for the love and support of my siblings and their spouses, Annie and Doug, Jeanne and Jeff, and John and Alisa. Most of all, I would like to thank my husband Daniel, who has taught me to know the actual ocean and to respect it. His wisdom, wide-ranging knowledge, lively curiosity, practicality, sense of humour, zest for life, excellent cooking, skilful IT support and fathomless love made this book possible.

PHOTO ACKNOWLEDGEMENTS

The author and the publishers wish to express their thanks to the below sources of illustrative material and/or permission to reproduce it:

Alamy: pp. 73 (Paul Fearn), 208 (Chronicle); Courtesy of the Beinecke Library: p. 146; Biodiversity Heritage Library: p. 94; Photograph courtesy of the Ove Christiansen Collection: p. 144; Courtesy of The Freshwater and Marine Image Bank, University of Washington Libraries: pp. 131, 140; General Motors: 205; Getty Images: pp. 166 (sspl), 179 (Bettmann), 196 (Bob Landry/ *Sports Illustrated*); Greenland Museum & Archive: p. 55; © Hans Hillewaert: p.218; Daniel Hornstein: p. 46; Indianapolis Museum of Art (Martha Delzell Memorial Fund): p. 127; iStockphoto: p. 226 (grandriver); Library of Congress, Washington, DC: pp. 78, 113; Mary Evans Picture Library: p. 150 (sz Photo/ Scheri); Courtesy of Terry Maas: p. 194; Courtesy of Maersk: p. 153; Mote Marine Laboratory: p. 198; Mohammed ali Moussa: p. 21; © Mystic Seaport Museum: pp. 100, 120; Photograph © 2018 Museum of Fine Arts, Boston: p. 126; nasa: p. 217; © Amgueddfa Genedlaethol Cymru/National Museum of Wales: p. 19; Nature Picture Library: p. 221 (Georgette Douwa); nnb: p. 49; noaa-hurl Archives: p. 26 top left & centre right; courtesy of Polynesian Voyaging Society and Ōiwi tv: p. 62; Klaus Bürgle and www.retro-futurismus.de: p. 164; Courtesy of Reynolds Brands: p. 174; Rijsmuseum, Amsterdam: p. 65; Shutterstock: p. 6 (Somkiat Insawa); Image by Pierre Thiret and Ann Randal and used courtesy of Andrea Sáenz-Arroyo and Comunidad y Biodiversidad, A.C: p. 224; un Photo: p. 185 (es); u.s. Federal Government (National Oceanic and Atmospheric Administration): p. 199; u.s. Fish and Wildlife Service: p. 201; University of California, Santa Barbara, Department of Special Collections, UCSB Library and courtesy of Get Oil Out!: p. 212; University of Washington Libraries, Special Collections (uw5647): p. 135; Vmenkov: p. 67; Wellcome Collection: p. 87.

INDEX

military 51
recreation 11, 92, 165–6, *166*, 191–7, *196*,
    202–3, 205–6, 219
research on 175–6, 180–1, 203
saturation 165, 174–81
and science 192, 194, 197–9, *198*, *199*, 202
subsistence 38, 65–6, 191
training and manuals 192, 195–7, 203–4
work 92, *164*, 165–6, *174*, 174–8, 180–81,
    189, 191, 194
Doggerland 41
dolphins and porpoises 19, 32, 35, 79, 170–71,
    179, 189, 195, 206–8, *208*, 216, 220
Doré, Gustave (illustrator) 126
Drake, Francis 88, 97
dredging 117–19, 121, 127, 173, 210
dumping waste at sea 210–11
Dutch East India Company 89, 94

Earle, Sylvia 198
East India Company 89
ecological restoration 223
ecosystem effects of marine exploitation
    36–7, 94–5, 132–3, 154, 217–18, 220,
    224
eels 26–7, 202
El Niño 223
energy from sea 166–7, 172, 219, 226
engineering 91, 105, 112, 161–5, 167, 175, 181,
    187, 192, 219
England 86–91, 96–7, 101, 107, 117, 136–8, *137*,
    195, 210
    see also Britain
environmental movement 209–12, *212*,
    215–16, *217*, 219, 221–3, 225, 227
Erikson, Leif 53
ethic, sea 225
Eurasia 18, 22, 29
Europe 77, 79, 81
    cultural discovery of sea 114, 117, 121–4,
        189
    exploration of Pacific 59–60, 64, 70, 93–8
    geographic discovery of sea 8, 10, 69,
        71–3, 75–8, 86, 89, 98, 102, 215, 226
    maritime culture 40–41
    maritime trade 83, 86–9
    naval power 88–9
    reliance on indigenous ocean knowledge
        74

reliance on marine resources 51, 53–4, 56,
    86–8, 132, 134, 136–42, 155–7, 213
struggle over free seas 89–91, 95–6, 103
voyaging and navigating 52–4, 106
evolution of life 14–15, 225
evolutionary time 9, 13, 39
Exclusive Economic Zones 11, 185–7
exploration
    Chinese maritime voyages 67, 67–9
    European, of ocean routes 69, 71–3,
        76–8
    and literature 122, 125
    of ocean versus space 165–6, 178–9, 182,
        188, 215
    of Pacific by westerners 59, 70, 93, 96
    of sea vs of land 54, 105, 107, 129
    personal 117–18, 162, 188, 191, 195, 197,
        202, 204
    reliance on indigenous knowledge 74
    scientific 29, 86, 98, 109–10, 163, 192
extirpation or extinction of species 20, 22,
    37, *94*, 94–5, 107
extreme environments 22, 24–5, 114, 205

Fairfree (shipping vessel) 155–6
farming the sea 167, 169–71, 182–3, 188
Fernández-Armesto, Felipe 39
Field, Cyrus 112–13
film industry, underwater 11, 177–8, 188–90,
    197–8, 203–4, 206–7, 222
Finley, Carmel 160
First World War 10, 141, 151–2, 158, 168
fish
    albacore 134
    cartilaginous 17, 19, 22
    effects of pollution and noise on 210–11,
        220
    haddock 53
    halibut 133–4, 138
    herring 53, 53, 141, 144, 156–7, 159
    jawless 17
    menhaden 134–5, 249
    plaice 136, 141
    salmon 27, 35, 53, 64, 120, 132, 134–5, *135*,
        138, 157–8, 160, 168–9, 187
    sardines 134
    scientific study of 190, 197–8, *198*
    targets for aquaculture 169
    tuna 48, 50, 134, 156–8, 187, 216, 223